THE CHANGING NORTH:

RECOLLECTIONS OF AN EARLY ENVIRONMENTALIST

JACK GRAINGE

About the author

Imagine living through an arctic winter in shelter with no plumbing, not even an outhouse. For two months, the sun does not rise above the horizon, and the temperature locks in below minus fifty. This is the story of Jack Grainge, a civil engineer for the federal government, bringing the people of northern Canada the sanitary conveniences we take for granted.

This was a difficult task. We think of great engineering works as gigantic projects constructed under daunting conditions. However, the construction of basic water and sewer services in areas where the ground is frozen to depths as much as several hundred meters, also required engineering ingenuity. Jack had a knack for finding simple, practical solutions to the problems. With a few notable exceptions, especially in later years, he was able to convince officials in other departments of the value of his recommendations. He also showed diplomatic skill in dealing with bureaucratic obstacles and the difficulties of cultural clashes.

Jack Grainge

Jack is a World War II veteran who took advantage of veterans' benefits to go to the University of Alberta and gain a degree in civil engineering. After graduation, he took further courses at the U. of A., and later gained a Master's degree from the University of California at Berkeley. Jack was often honored by his peers for his research and other accomplishments. He received two travel fellowships from the World Health Organization (WHO) to study community planning and municipal engineering in northern Scandinavia, Greenland, and Iceland. The California Chapter of the Society of Sigma XI awarded him an associate membership. The Professional Institute of the Public Service of Canada awarded him a gold medal, and the Northwest Territories Association of Professional Engineers, Geologists, and Geophysicists presented him with an Honorary Member Award.

John Shaw
John Shaw, an employee of the Alberta Department of the Environment, spent his early engineering years in Jack Grainge's office. He also served a while as Jack's successor.

THE CHANGING NORTH:

RECOLLECTIONS OF AN EARLY ENVIRONMENTALIST

JACK GRAINGE

Occasional Publication No. 47

Canadian Circumpolar Institute
University of Alberta

Canadian Cataloguing in Publication data

Grainge, J.W.
The changing north

(Occasional publication series, ISSN 0068-0303; no. 47)
ISBN 1-896445-15-2

1. Grainge, J.W. 2. Sanitary engineering, Low temperature. 3. Sanitary Engineering—Canada, Northern. 4. Water supply engineering—Canada, Northern. 5 Civil engineers—Canada, Northern—Biography. I. Title.
II. Series: Occasional publication series (Canadian Circumpolar Institute.
TD940.G72 1999 628'.097192 C99-910957-X.

ISBN 1-896445-15-2

Cover design by Graphic Design Services, University of Alberta
Printed by Quality Color Press Inc.

This publication was made possible with support from the Government of the Northwest Territories, Department of Municipal and Community Affairs.

INTRODUCTION

Previous to World War II, the Canadian North was a largely unknown frontier. The common perception was of a land of ice and snow, occupied by subsistence hunters and gatherers. Unfortunately, this perception was largely true.

The Hudson's Bay Co. trading posts were the most prominent feature at the main gathering centers. Nearby were Catholic and Protestant missions, Royal Canadian Mounted Police Stations, and Department of National Health and Welfare Buildings. In the western Northwest Territories, there were church-operated hospitals and residential schools at Aklavik, Fort Simpson, and Fort Providence and church-operated hospitals at Rae, Fort Resolution and Fort Smith. During WWII, the American Army built a chain of airports from Edmonton to Norman Wells.

In 1950, Prime Minister Diefenbaker made an historic visit to the western Arctic. He found a need for a greater government presence in the North. He was dismayed when he found that people in Old Crow, Yukon Territory, received mail by way of Alaska. The residents mailed letters bearing American stamps by way of Alaskan airlines. He also noted the dearth of health services available to the native population throughout the North. His positive response was reflected in his position paper, Vision of the North.

More schools and nursing stations were built at remote settlements. Inuvik, a major administrative and education center, was constructed with piped water and sewer services. Many other towns were built where there were formerly small settlements. To plan the water supplies, sewers, and wastes treatment and disposal for some of these communities, the Department of Indian and Northern Affairs hired southern-based, consulting engineers. Jack Grainge, with an understanding of the problems, provided useful advice to both government-employed and consulting engineers. In addition, Jack Grainge, together with his engineers, participated in the education of, and advice to, local community health workers.

Jack Grainge measured up to many difficult challenges. I traveled with him on many occasions and witnessed at first hand, his dedication and ability to cope. It is my sincere wish that his efforts will receive the respect and honor that they deserve.

Respectfully submitted,
N.A. Lawrence, P. Eng.

PREFACE

In the mid-1940s the citizens of Yellowknife and Aklavik installed surface water distribution systems to supply water to their houses during the summer. This was a luxury for them, but for the rest of the year they had to haul or carry water to their homes. In the rest of the communities of the Northwest Territories even a summer-only piped water system was a "pipe dream." In most of the communities, the people threw waste water on the ground, near their doorways and discarded toilet wastes and garbage a distance away to be disposed of by gulls, ravens and scavenging dogs. In a few larger communities, toilet wastes and garbage were hauled to isolated places nearby. Such were water, sewage and garbage services in the Northwest Territories before 1949.

In 1950 I graduated as a civil engineer from the University of Alberta. I chose, as a career, water and sewage engineering, and stream, air and land pollution protection. The regional engineer of the Public Health Engineering Division (PHED) of the Department of National Health and Welfare (DNH&W), based in Edmonton, offered me a position in that field that was soon to open.

While awaiting that job opening I worked three months for Associated Engineering Services Ltd. (AESL), Edmonton, checking the construction of water and sewerage systems in St. Paul, Alberta. I then worked temporarily in the Department of Public Works of Canada. When the position in PHED, for which I had been waiting, became vacant, I transferred to that office.

What was my new job? I became assistant to Stan Copp, the region engineer in a two-man office in Edmonton. Our territory was Alberta and the western Northwest Territories. We were required to recommend compliance with sanitation and environment requirements on all properties under federal jurisdiction. This included the Northwest Territories (NWT), national parks, Indian reserves, railway properties, airports, and so on. In the NWT we worked with the public health doctors and nurses in the DNH&W, and sent recommendations to the PHED head office in Ottawa. Usually they forwarded our NWT reports to the head offices of the Department of Indian and Northern Affairs (DI&NA) or the Department of Transport (DOT).

When the Government of the Northwest Territories came into being in 1967, I worked with the territorial engineers in Yellowknife. However, my reports went to them by way of my head office and the Department of Indian Affairs and Northern Development, the new name for DI&NA.

In the early 1970s, our office was absorbed by the newly organized federal Department of the Environment (DOE). In 1975, I transferred back to DNH&W. Both territorial and DOE offices in Yellowknife were concerned in that field.

In 1994, seven years after I retired, Vern Christenson, Assistant Deputy Minister of Municipal and Community Affairs, Government of the Northwest Territories, suggested that I write my memoirs. However my memory has been failing fast. I have done my best to ensure that information outstrips my imagination.

Unfortunately I did not keep a diary. I did not even save field notes. I left my files in the office for John Shaw, who succeeded me. After he resigned, they either became buried in old files or were trashed. Fortunately, Ev Carefoot, a retired engineer of AESL, had saved a few of my 1969 and 1970 reports. I had mailed them to him when he was making reports of his own.

I am grateful to Vern Christensen, Ron Kent and Terry Brookes, my contacts in the NWT administration, for their encouragement and advice. Also I appreciate the excellent suggestions by the reviewers of my submission to the Canadian Circumpolar Institute. In my writing, I referred to the publications Water and Sanitation Services, Northwest Territories, November 1981 and NWT Data Sheets. At the end of each community report, I listed other publications to which I have referred and the names of friends who provided information.

I am indebted to many people in the communities in the NWT, who helped me with my work, by answering my torrents of questions. The Inuit and Dene, the native people of the NWT, were always friendly and helpful. Many of them welcomed me in their homes even when we could not speak a common language. Their delightful gestures and smiles expressed their irrepressible humor. We had many hilarious moments together.

I am grateful to the "outsiders" who worked in the North. I found them to be generally dedicated. Thanks to those who helped and accommodated me.

Doctors, nurses and administrators in the DNH&W cheerfully accomplished their challenging duties. They and my own staff, made useful contributions. Elsie Stannard, my stenographer and receptionist for six years, was particularly friendly, efficient and hardworking. She still congratulates me on my birthdays.

I am indebted to many people who checked my writings. John Shaw, and Mike Pich, formerly on my staff, University of Alberta Professor Emeritus Jack Bilsland and members of his writing class to which I belong, and my wife Jess. Also my thanks to Jim Cameron, consulting engineer, Bob Milburn, who held several senior engineering positions in the Government of the NWT and the City of Yellowknife, Brian Edwards, whom I had hired to do research and later went on to direct a large research staff and Vern Christensen who reviewed the whole manuscript and offered useful guidance.

I arranged the chapters in this book beginning with Kugluktuk (Coppermine), one of the first communities. Next follow the other Arctic Coast communities, then those along the Mackenzie River and finally those further south.

CONTENTS

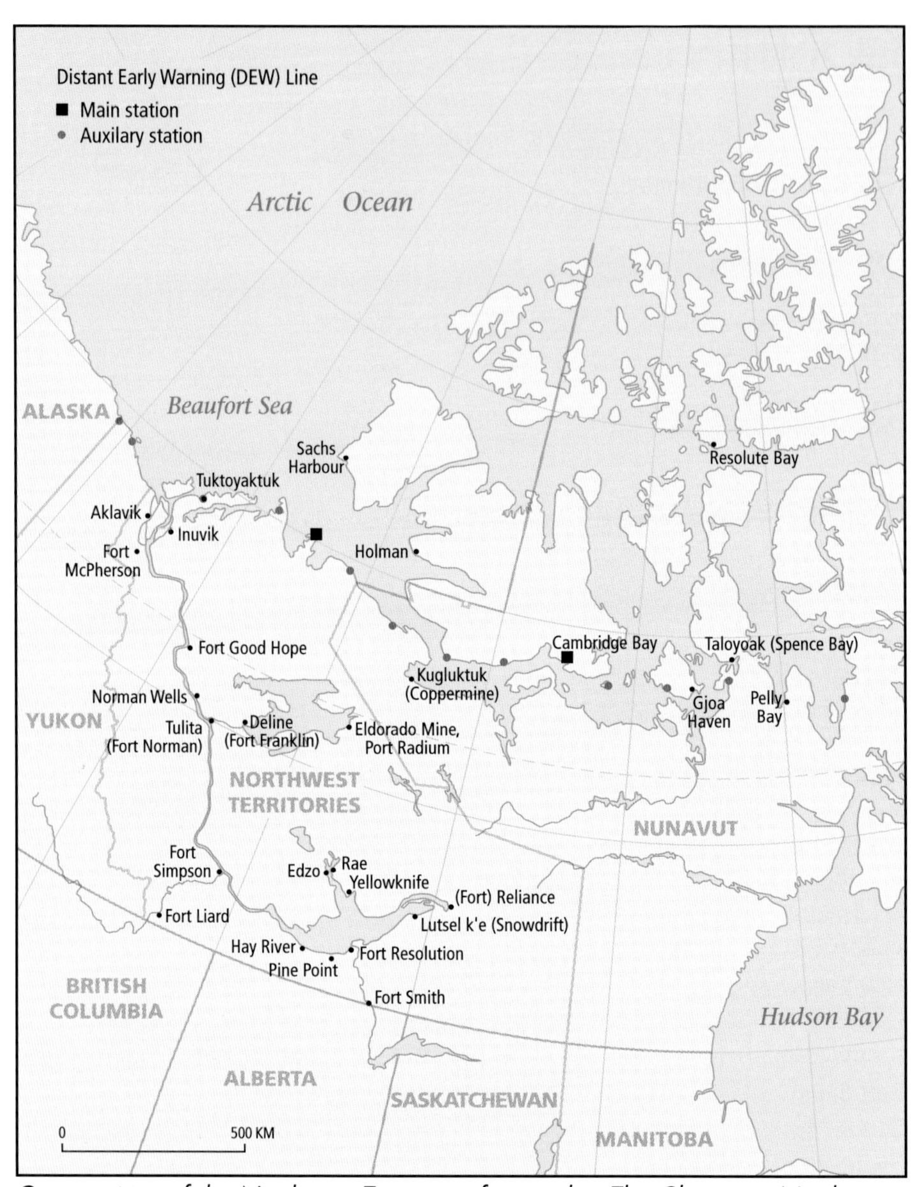

Communities of the Northwest Territories featured in The Changing North, showing Distant Early Warning (DEW) line main and auxiliary stations. Adapted from EPEC Consulting Western Ltd. (1981), by Johnson Cartographics Inc., Edmonton.

ELDORADO MINE, PORT RADIUM

Eldorado Mine was built on the steep-cliff shore of a small peninsula, projecting into Echo Bay, in the north-east corner of Great Bear Lake. The land is mainly bare rock, with broken rocks, soil, and lakes inland. The buildings of both the mill, offices, and residences were white with red roofs. When I landed there on a sunny winter day in April 1953, it was an eclectic mixture of buildings in tiers. It was not beautiful, but it was aesthetically pleasing.

On the high, flat land behind the camp, the crews had built a tennis court, a football field, and a one-sheet curling rink. There was an inexhaustible number of large land-locked salmon to be hooked. The camp, protected from the fierce north winds by the south-facing cliffs on which it was built, overlooked the beautiful lake. Home-made brew was plentiful. The parties, especially the shivarees, were high spirited. No wonder moral was high.

* * * *

Gilbert LaBine was a solid, broad-shouldered man of medium height. Of Irish and French-Canadian parentage, he was a good student. In 1905, at age fifteen, he quit school and went to work. Later his tall, husky, older brother, Charles, and he went prospecting in Ontario and south-east Manitoba. During the winters, Gilbert studied geology. Their prospecting was moderately successful, and Gilbert organized a company, Eldorado Gold Mines, Ltd.

In 1900, geologist MacIntosh Bell, of the Geological Survey of Canada, and his young helper, Charles Camsell,

Figure 1. *Eldorado Mine, Port Radium (1955). Photo by Jack Grainge.*

who was born in Fort Liard, surveyed the area around Hunter Bay in the McTavish Arm of Great Bear Lake. He found quartzite with indications of copper and gold.

In 1929, an early bush pilot, Leigh Brintnell, flew Gilbert LaBine into the same area and extended his search southward. On his return flight to Edmonton, while flying over Echo Bay, LaBine noted rocks of many different colors.

The following spring, LaBine returned with Charlie St. Paul. He ran across veins of a black thread-like substance, and staked several claims. Meanwhile, Gilbert's brother, Charlie, and his men, with food and equipment, in large motorized canoes, sailed up the Athabasca, Slave, Mackenzie, and Great Bear Rivers to Echo Bay. LaBine staked many more claims and the men constructed log buildings, a staff house and warehouse. Not having a circular saw, they whip-sawed the lumber for the floors and roofs. In early winter they flew south.

Figure 2. *A second view of Eldorado, showing its setting within the bay. Photo by Jack Grainge.*

The samples they brought back that fall proved to be rich in pitchblende, as well as copper, cobalt, silver, and bismuth. The pitchblende contained radium. The Belgian Congo was now no longer the only source of costly radium. Gilbert LaBine and his company were rich. He named the mine, Eldorado Mine Ltd., and the place, Port Radium.

The samples also contained uranium, for which there was little demand. European firms used it sparingly as an additive for the coloring of glass, pottery, and enamelware. It improved the finish of the glaze.

In 1931, Gilbert LaBine brought in crews to develop the mine and build quarters for the workmen. When production began, they washed the uranium salts out with the crushed residue into the tailings pond behind the rocks back of the mine. The Mackenzie River Transport hauled the equipment in and the bags of ore out. Later that year, LaBine transferred his shipping business to Northern Transportation Co. Ltd. Eventually, he bought that company.

During succeeding years, bush pilot Stan McMillan, flying a Bellanca plane hauling a capacity load of two tons, brought workmen and supplies to the mine. On each return trip to Edmonton, he carried a full load of ore. In winter, he landed on skis at McCall Field, now called the Edmonton Airport. In summer, he used pontoons, landing on Cooking Lake, southeast of Edmonton.

* * * *

On June 11, 1941, the federal war-time, manpower office ordered Eldorado Gold Mine to close. Gold was unimportant for the prosecution of the war. Mine Manager, Ed Bolger, shut it down. When the de-watering pumps were turned off, the water in the mine rose from the 275 m bottom level to the nine meter level, the same as that of Great Bear Lake.

On April 21, 1942, the Hon. C.D. Howe, at fair market value, purchased for the Government of Canada, Eldorado Gold Mines Ltd., including the subsidiary company Northern Canada Transportation Co. Ltd.

Howe ordered Gilbert LaBine to immediately open the mine and increase production to the maximum. The work was secret. Howe ordered LaBine to tell no one, not even his wife. The miners hired were screened by the RCMP and sworn to secrecy. In order that orders for equipment and transportation be given the highest priority, they were marked "Manhattan Project." Since gold had no priority with suppliers and shipping companies, the company name was changed from Eldorado Gold Mines Ltd. to Eldorado Mining and Refining Ltd., LaBine ordered Ed Bolger to quickly reopen the mine and begin the shipment of the concentrate to Port Hope, Ontario.

He asked pilot A.B. (Alf) Caywood to organize the newly-formed Eldorado Aviation Division of the mining company. Prior to the shut-down in 1940 of Eldorado Mine, Caywood had been flying a Canadian Airways plane supplying that mine.

The mill processed both the ore that was mined and the tailings containing uranium salts, which had previously been discarded.

* * * *

About five o'clock one morning in April of 1953, I climbed into the belly of a DC3 plane at Edmonton. Several other passengers and me, all wearing parkas and overshoes, sat side-by-side on a bench along the lee-side of the plane. Boxes of supplies were ahead of the doorway on the opposite side. The famous bush pilot, Alf Caywood, scrambled on board. He seemed like an expressionless, ordinary workman, not my perception of what the senior manager of the Eldorado Aviation Division should be.

At Eldorado, the affable Harold Lake welcomed me and took me to the guest suite. I ate at the clean, single-men's mess hall. Both the accommodation and the meals were excellent. Fresh vegetables, fresh milk, cleanliness—no complaints.

Water for both domestic and mine use was pumped from the Bay in front of the settlement, but I cannot recall any details. The sewage, garbage, and refuse were discharged to the back side of the peninsula. The mine tailings were discharged over the bank near the end of the peninsula.

In 1954, Harold Lake moved south to become Assistant Manager, and later Manager, of Eldorado's huge operations at Beaverlodge, in northeast Saskatchewan. A year or two later, due to cancer, he retired in Edmonton. He joined the Royal Glenora Club where he and I played several games of tennis together. He had played tennis on a court at Eldorado, Port Radium.

John (Jock) McNiven replaced Lake at Eldorado, Port Radium. He had been the manager of Negus Mine in Yellowknife. In that mine's dying days, he won an election as Yellowknife's Mayor. Jock could let out a stream of unprintable oaths, but displayed a heart of gold. No wonder everyone who worked for him, or even knew him, loved him.

In early 1960, the ore deposits were dwindling, so the crew had diminished. No longer were there games of hockey, baseball, and curling. However, McNiven developed a rapport with his men. He tried to make camp life interesting. Once he hired Al Oeming, Edmonton's popular mammalogist and wrestling promoter, to bring in six male wrestlers and two female wrestlers. McNiven insisted on the ladies' participance. He knew men's interests.

The mine 'bottomed out' at 427 m. Both the ore and the tailings that they were reprocessing, gave out. McNiven closed the mine on September 16, 1960. He came to Edmonton, stayed for awhile at the Corona Hotel, where I went to visit him. He retired to Saltspring Island in British Columbia.

Echo Bay Mines bought the property and claims and are now operating mines in the vicinity.

Reference
Lake, H. 1953 and in Edmonton approximately 1965.

KUGLUKTUK (COPPERMINE)

Kugluktuk is situated on the south shore of Coronation Gulf, a few meters west of the mouth of the Coppermine River. There is good fishing and sealing off the coast in the vicinity. Fish are abundant and easy to spear in shallow water below the falls, nineteen kilometers upstream. Sometimes a caribou herd migrates within eighty kilometers to the south. Considering the good hunting and fishing potential, Inuit have probably lived in the vicinity for millennia.

In 1771 Samuel Hearne, with the assistance of the great Chief Matonabbee and his party of Yellowknife Dene, walked across trackless tundra and then down the Coppermine River to its mouth. The crew killed a few Inuit who were fishing below the falls. Being their deadly enemies at that time, perhaps they thought it was a case of either kill or be killed. Hearne claimed the area in the name of the HBC and named the river after the surface copper mine. The copper was a resource long known to the Inuit. Fifty years later Sir John Franklin explored the coast of the Arctic seas from the Mackenzie River to 280 kilometers east of the mouth of the Coppermine River. In 1865, whalers transmitted influenza to the Inuit, causing thirty percent mortality. In 1912 Catholic Fathers Rouvière and Leroux came to Kugluktuk to establish a mission. Perhaps feeling threatened, the Inuit murdered them. In the years 1913 to 1916, ethnologist Diamond Jenness studied the culture of the Inuit in the vicinity.

Figure 3. *Approaching Coppermine from the air. Photo by Jack Grainge.*

In 1916, free trader Charles Klengenberg set up a trading post at the present site of Kugluktuk. Eleven years later the HBC established a trading post and Klengenberg moved to Cambridge Bay. In 1928 Canon Webster established an Anglican mission. A year later Dr. Russel opened a temporary hospital to help treat an influx of survivors of an influenza outbreak at Bernard Harbour, one hundred kilometers to the north. In 1932 an RCMP post was established, followed in 1937 by weather and radio communication stations, in 1948 by a nursing station with Mrs. Dufresne, as nurse, and in 1950 by a one-room school.

Doug Lord became the first school teacher, and Cathy, his wife was the nurse. After two years in Kugluktuk and a year in Fort Simpson, Lord transferred to the patient teaching staff at the Charles Camsell Hospital in Edmonton. That hospital, organized at the end of World War II, served Dene and Inuit. To relieve the boredom of long-term care, the staff provided the tubercular patients with handicraft materials. The patients enjoyed those crafts. Lord taught mathematics and helped with the carving program. His collection of Inuit art is an important part of the Alberta archives. At Kugluktuk he had introduced the Copper Inuit to the profitable art of soapstone carving. Lord, whose hobbies included building toys, furniture and radio-controlled model airplanes, was, in both Kugluktuk and the Camsell Hospital, the right man in the right places at the right times.

During World War II, while we were both serving on an RCAF bomber station at Leeming in Yorkshire, Doug Lord and I became friends. After the war, we both attended the University of Alberta, and graduated in 1950.

In late June 1953, I flew to Kugluktuk in a single-engine Norseman to study the water and wastes systems and make suggestions for their future development. The pilot was Ernie Boffa, a famous bush pilot who is listed in Canada's Aviation Hall of Fame. During the early part of World War II, he had been a pilot of Canadian Airways Ltd. flying for RCAF navigation students, including the young Stanley Reynolds. After the war Reynolds created the famous Reynolds Museum in Wetaskiwin, Alberta.

In 1943 the Eldorado uranium mine on Great Bear Lake was a wartime priority and needed an intrepid, dependable pilot to make regular trips there. Canadian Airways transferred Boffa there to fill that need. Unfavorable weather could delay him but not stop him. If when he was flying, the weather socked in, Boffa would wait for flyable weather on a convenient lake. He carried a sextant to determine his location and thereby never became lost. Also Boffa could repair his plane, and was the dependable pilot needed for supplying Eldorado Mine, the source of uranium for the atomic bombs.

Betty, the wife of Canon Jack Sperry, was also a passenger. She was returning with her first new-born. The Reverend Sperry later became bishop of the Diocese of the Arctic.

At Port Radium, we picked up the Eldorado Mine doctor, an amiable, stocky Briton, about thirty-five years of age. He was making his once-a-month visit to the nursing station at Kugluktuk.

At take off from Port Radium, Boffa demonstrated his resourcefulness. The plane's starter would not work. He hooked a loop at the end of a long rope over one end of the propeller and wound the rope several times around the hub of the propeller. Then about eight men from the mine, as well as the doctor and I, ran along a path away from the plane pulling the rope and thereby twirling the propeller. After the last wrap of the rope unwound, the loop slipped off the propeller and we human dominoes toppled over. After a couple of tries, the engine started and we were away.

* * * *

At Kugluktuk, Boffa landed at a raft anchored to the seashore fronting the HBC trading post. The post consisted of a house, a store and a warehouse, all of them white with red shingle roofs, typical HBC colors. The post was near the wharf. The Anglican and Catholic missions were slightly farther inland, and well spaced on either side of the HBC. The nursing station was south of the Anglican Mission and about two hundred meters from the shore. Farther west was a one-room school with a teacher's suite on the second floor. The other government residences and offices were widely spaced in a row to the west, about fifty meters inland. Counting the new baby, there were ten outsiders. About twelve Inuit lived in tents and small huts to the east. The graveyard was on a nearby Delta island.

Ours was the first plane to land on the water that year, five weeks after the last plane to land on the ice. During the break-up of the ice and for a few days thereafter, the water was choked with ice chunks and therefore not safe for landing.

All the outsiders turned out for the occasion, teacher, nurse, the Department of Transport (DOT) weather observer, the DOT radio signal man, the HBC manager and his assistant, Catholic Fathers Lapointe and Adam and Anglican Canon Sperry. The RCMP constable was away on a routine visit to fishing camps. There were about six Inuit, who had turned out to help the HBC manager and his assistant unload supplies.

Miss Kitchen, a quiet, friendly nurse, invited the doctor and me to dinner at the nursing station, a wood-frame bungalow with a gable roof. It consisted of a waiting room, an office-patient examination room, two wards, a kitchen-dining room, bedroom and bathroom.

She prepared an unforgettable, delicious dinner of fried arctic char, and she seemed to be glad of our company. She had been hired in England. On her way to her job she saw little of Canada, other than the Charles Camsell Hospital at Edmonton, her headquarters. She said that she enjoyed Kugluktuk's peacefulness, but the winters were depressingly long. The doctor thought that she should return to city life. She said she had cousins in Calgary, Ralph and Ellen Clark. By coincidence when I lived in Calgary during the 1930s, I knew her aunt, uncle and both cousins. Ellen had been my first girlfriend.

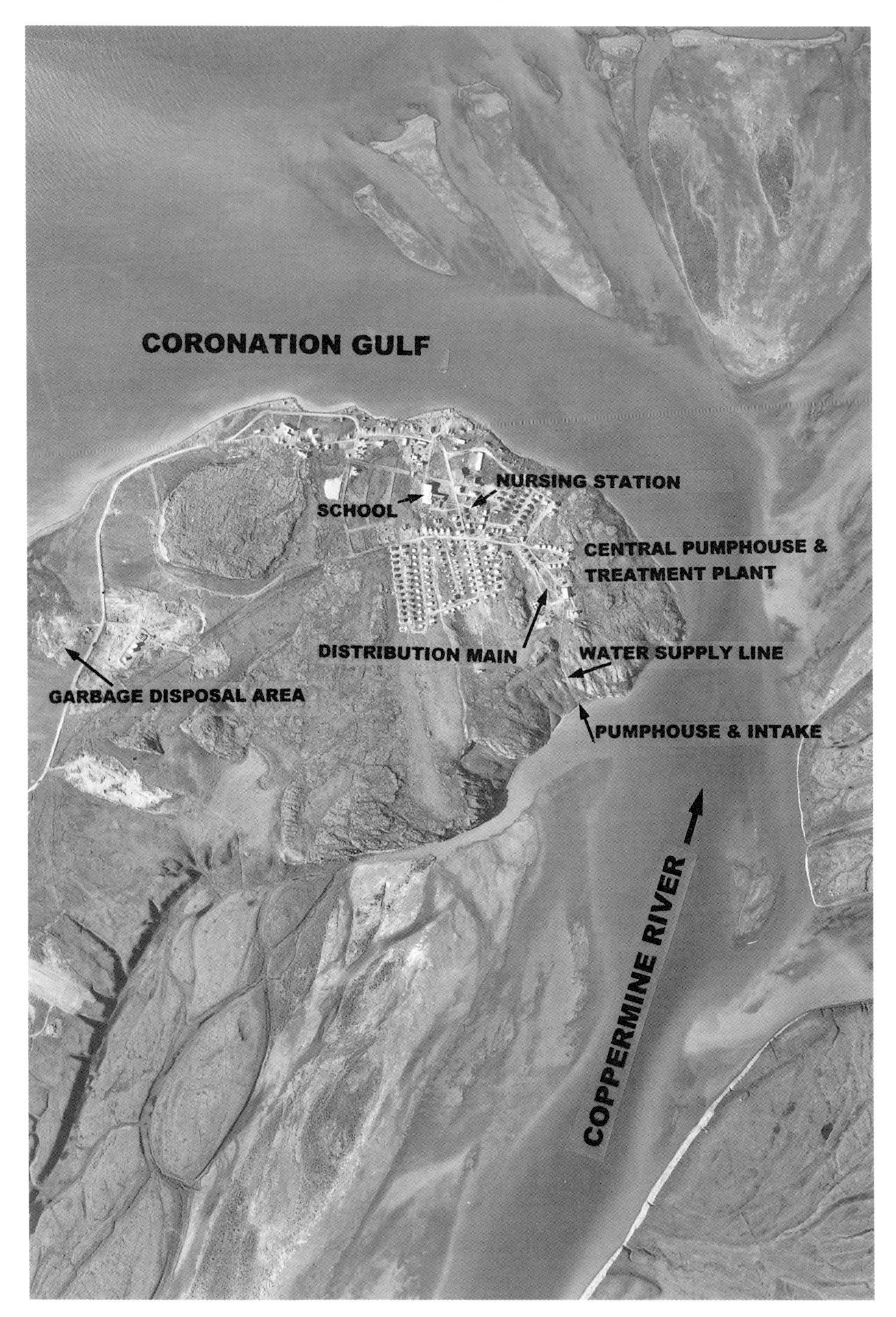

Figure 4. *Aerial photo of Kugluktuk. Department of Energy, Mines, and Resources, 1979.*

For isolated communities such as Coppermine, the medical health officers whenever possible recruited British nurses with midwifery training. In time I learned that the nurses have been under-recognized for their invaluable contributions to public health in the North.

After dinner the doctor and I began a walk around the settlement. We went first to the house of the weatherman. The doctor accepted an offer of a drink. That ended his walk. I went on asking questions of others in the community.

While I was talking to Leo Manning, the HBC manager, I was looking at the ocean beyond Graveyard Island in the delta. All of a sudden I witnessed a mirage. High, white cliffs rose instantaneously out of the sea. Manning said that there was no land there.

* * * *

The water supply posed a difficult problem. The river water was clear and soft, but it was sometimes difficult to reach. People could not walk to the river from the settlement because of a wide strip of high sharp rocks near the shore. In summer some people motor boated along the seashore and a short distance up the river to obtain clear water.

In winter, after ice had formed on the sea, river water displaced the salt water. Thus, people could chip a water hole through the ice on the sea in front of the settlement. During the spring and fall, the people depended on ice blocks that they had cut on the river. In winter they stored the blocks outside, and at other times kept them packed in sawdust in ice houses. When someone wanted more water, he washed another ice block and dumped it into his water barrel (washed oil-barrel). Miss Kitchen used ice throughout the year. People dipped water from their water barrels, somewhat unsanitary, but not unlike common practice at that time on farms and in small communities throughout Canada.

As a boy during the depression in the 1930s, I lived for a few years on a homestead in northern Alberta. For one of those years I lived in a crowded, small log cabin. I drew water from a well with a bucket. If a drowned mouse was in my bucket, I considered that water to be contaminated. I poured it into the trough for the horses and cattle. If succeeding buckets did not have mice in them, to me that water was uncontaminated. I carried it to the house. Everyone, including visitors drank water from the bucket with our dipper.

In our one-room, eight-grade, log-cabin school, the teacher would explain that people should use separate glasses for drinking, but we had only one dipper for her and all of the students. Every other Friday was dance night, and everyone in the crowded room drank water from the same dipper. Under the circumstances, that was the only practical method.

The Inuit employees emptied the honey buckets and dumped garbage in out-of-the-way places where scavenging dogs and birds would dispose of them. At the time this was common practice in northern Canada, northern and western

Alaska and Greenland. Outsiders threw their wastewater a short distance from their houses. That practice was not sanitary but at the time people on farms and in small communities in Alberta were doing the same, but in their own yards.

From the door stoop on our homestead we threw liquid wastes as far as we could fling them. On dance nights at our school, I know that the men never walked as far as the outhouse.

While the people in arctic Canada were living nomadic lives, it was not so necessary for the people to follow sanitation rules. In any case, modern sanitation practices were virtually impossible in cramped, unbelievably small snow houses or tents. The people lived in balance with nature, usually traveling in one, two or three family groups. According to the early public health nurses that I met, usually children were not born within about four years of one another. Many of the nurses and doctors thought the gap was due to mothers suckling their babies until they were three or four-year-old toddlers.

* * * *

Some people complained that the level plain on the opposite side of the river was sunny for many weeks in winter while Kugluktuk, being at the base of a high, north-facing hill, was shaded. However, despite legions of mosquitoes, sometimes dirty drinking water, ugly honey buckets and long, cold, dark winters, everyone professed to like living there.

I bought some Inuit soapstone carvings from the teacher. It was the first time I had ever seen or heard of such large soapstone carvings, ten to twenty centimeters long. I had friends living near Calgary, Geraldine and Fred Percival. He had inherited the title of Earl of Egmont. She had tiny soapstone carvings. Her intrepid grandfather, Inspector J.D. Moody, had collected them while serving in the North West Mounted Police during the Klondike gold rush. He told her that children in nomadic families could carry only small, lightweight dolls and toy animals. Doug Lord, the first teacher at Kugluktuk, had taught them how to make soapstone carvings.

When I returned from Coppermine to Edmonton, I decided to advertise the art for the benefit of the Inuit. I phoned an Edmonton Journal reporter who wrote an article and included a photograph of me holding the carvings. From then on friends asked me to bring back carvings for them.

* * * *

Meanwhile Boffa had fixed the plane's starter. I sat up front, where I discovered how Boffa could fly for hours and hours, day after day. He spent an hour or so with his head turned away from me, apparently looking out of his side window. With his arms resting on convenient places, he held the control wheel steady. Then he set up a rhythm of sleeping for two or three minutes and waking up for a minute or so. He woke up to point out to me a herd of musk oxen.

* * * *

In my report on Kugluktuk, I stated that obtaining fresh water was a problem because the river was not conveniently accessible. Also, the community was in the shade of a high hill. I recommended finding a better location for the community. So the Ottawa decision-makers decided that the survey team, that would be looking for a new site for Aklavik, should also look for a better site for Kugluktuk.

Thus in July 1954, the now-rich-and-famous Max Ward transported the survey crew from Aklavik to Kugluktuk and back in his new, single-engine, Otter aircraft. Curt Merrill, our leader, piloted the aircraft part way. Both he and Ward had been wartime pilots in the RCAF. The rest of the party were John Pihlainen, Hank Johnson and Roger Brown, of the permafrost section of the National Research Council, Keith Fraser of Geological Surveys, all of Ottawa. From Edmonton there was Ken Berry of the Department of Public Works, Ed Garret of the DOT, and myself.

Ward told us that he had won the contract to transport us because his single-engine Otter was the largest plane based in Yellowknife. It was the only one that could take our whole crew and our baggage from Aklavik to Kugluktuk and on to another possible site and back. He was the first airline operator to realize that there was a need for a larger aircraft than the Norseman that most Yellowknife airlines were using. His business acumen paid off because although he started his company late, he soon operated the largest airline in the North. Later Wardair, his company, became the third largest airline company in Canada.

Because the area between Aklavik and Coppermine was not mapped, we could not fly there directly. We flew a much longer distance, north to the coast and then east. In passing we looked down on the lonely Catholic mission at Paulatuk.

At Kugluktuk we ate and bunked at the school, having brought our own sleeping bags. The next morning we flew to Port Epworth at the mouth of Tree River, 140 kilometers farther east. No one lived there. Six Inuit, fishing along the opposite shore, paddled across in their wooden boat and greeted us with smiles. We explained our purpose to one of them, who had learned to speak English while a patient in the Charles Camsell Hospital in Edmonton.

We found no suitable site. Therefore, we returned the same day to Kugluktuk. The next day we looked around that site. John Pihlainen and his crew tested the soil with a meter-long, hand-driven auger. I favored a site across the Coppermine River from Kugluktuk, and Father Lapointe took me there in a boat. The others on the team considered that the river there was too shallow for a port.

We found a deep pocket in the river, immediately beside a rock shelf upstream of the highest tide level. It was over the hill and around the end of the ridge of high rocks that separated the river from the settlement. It could be an excellent pump site. Immediately we felt more optimistic about the present site. The buildings were already there. The land had a gentle slope, making it suitable for the operation of future gravity sewers. The view of the Gulf, the delta islands and a point of land to the west, was spectacular. Everyone agreed that it was a good site, but I thought it was too shaded. However, at the time I did not know of a better site.

We did not realize it, but the south side of the hill behind the community, overlooking the present airport, was a good site. In winter there would be about fifty fewer sunless days, but it would seem like many more. The hill would provide some protection from the cold, north winds. It was only a short distance from the new-found water intake on the Coppermine River. At the bottom of the hill to the southwest, there was a suitable sewage lagoon site.

However, at the time we did not think of these advantages. We were thinking that it would have meant a two kilometer-long road from the HBC dock. Considering the road construction equipment available then, that was a long distance. Belatedly I wish I had recommended the site on the south side of the hill. I think if I had done so, the planners in Ottawa would have accepted it.

Later that day I talked about the possibility of constructing water mains and sewers. Everyone laughed and I joined in. It sounded like a weird idea. The population of the community was only about twenty. Even the Ottawa administrators were not expecting the later huge in-migration of Inuit from outlying camps. They were thinking of bringing students to live in communal tents while going to school. An Inuit couple would look after the children in each tent. After the school year, the children would return to their parents in their hunting camps.

No one involved in the planning realized that the Inuit would not be separated from their children. The parents wanted to be with them in order to teach them the skills of hunting and trapping, traveling long distances with scarcely visible landmarks, making sledges, snow houses and warm clothing.

* * * *

On a visit to Kugluktuk ten years later, I met Les Clark, an engineer constructing DOT buildings on the western edge of the settlement. He was competent and he had imagination. He wanted to construct practical, piped water and sewer systems for two offices, three houses and a ten-man barracks.

We discussed a plan for water and sewer systems which he later constructed. Each building would have a water reservoir and a sewage tank, each vessel with a one-week capacity. To avoid sewage overflows, in each case the sewage tank was slightly larger than the water reservoir. Clark designed the systems.

In summer, a gas engine-driven pump discharged water from the river through a 4.5 kilometer-long, surface pipeline to the DOT group of buildings. The pipe was 30 mm, soft copper, with soldered joints. It snaked through a gap in the rocky area, and ran south of the community to the DOT group of houses. Clark chose the route so that it ran continuously downhill. The pipeline could then be drained after summer use. It remained in place throughout the winter.

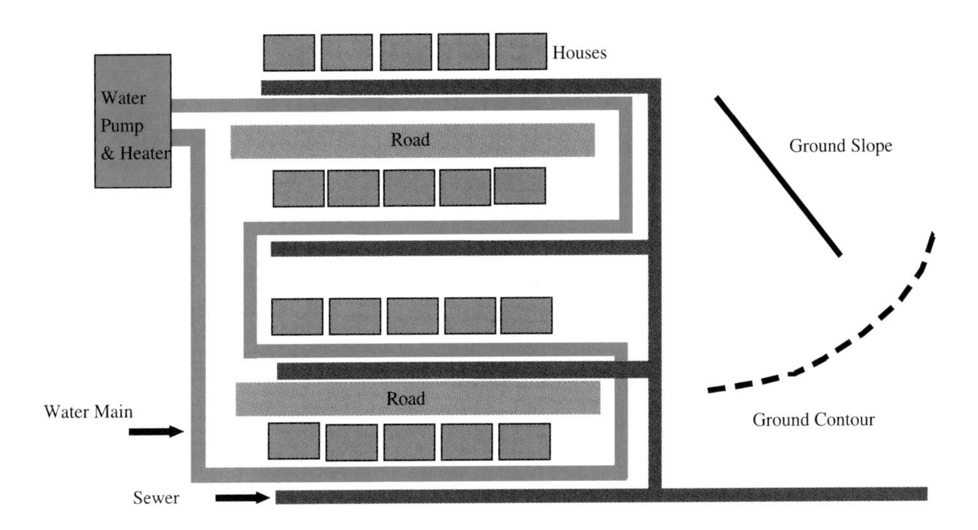

Figure 5. *Suggested water and sewer plan for northern communities, designed by Jack Grainge.*

In winter, a small shack was placed over a hole chipped through the sea ice near the shore. A gas engine-driven pump in a heated shack discharged water through a bare, aluminum pipe, laid on the ground surface. The water discharged into a reservoir in a building heated by cooling air from the power generators. Immediately after the pumping stopped, the operator stored the water pipe sections on end in the water reservoir building. The ice in the pipes thawed and the water drained away. The reservoir held almost a five-week supply of water. Thus, the operator could pump water during mild spells.

Once a week, a pump in the reservoir building discharged water through a water pipe to the household reservoirs. At the same time, household pumps discharged sewage tank contents to the sewer. The hard copper, water distribution pipe and steel sewers lay side by side on a cement-asbestos board, with a Pyrotenax, electric heating cable between them. These pipes, surrounded by rigid foam polyurethane insulation, lay in a 30 cm x 30 cm, galvanized iron utilidor. The top and bottom halves were identical and had flanged edges that were bolted together. This construction made it possible to open the utilidors in case repairs were required.

The utilidor ran from one building to the next, and continued to outfall at the sea shore thirty meters beyond the furthermost building. The utilidor was laid to grade so that both the water pipes and the sewers would empty after both the water and sewage flows had stopped. The operator switched on the electricity to the heating cable a half hour before the water and sewage pumping began. He turned it off a half hour after the pipes had drained. Valves in the houses automatically closed when the water reservoirs had filled.

During the summer, the RCMP connected their houses and office to the systems. Although the DOT administrators invited other departments to do the same, none of them took advantage of the generous offer. I suppose it would have meant too much work to install taps in their kitchens.

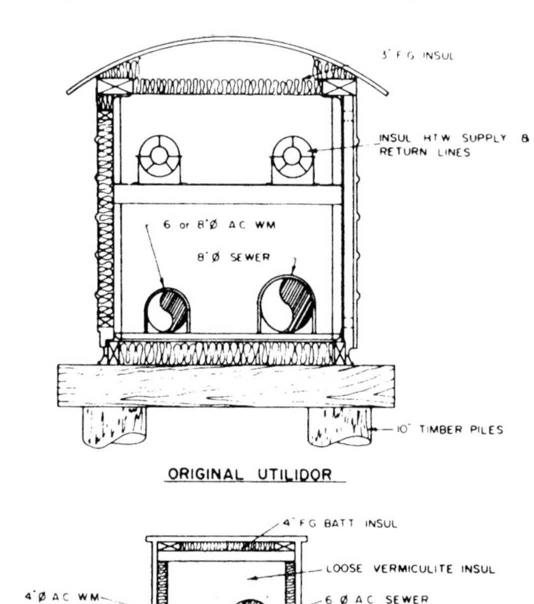

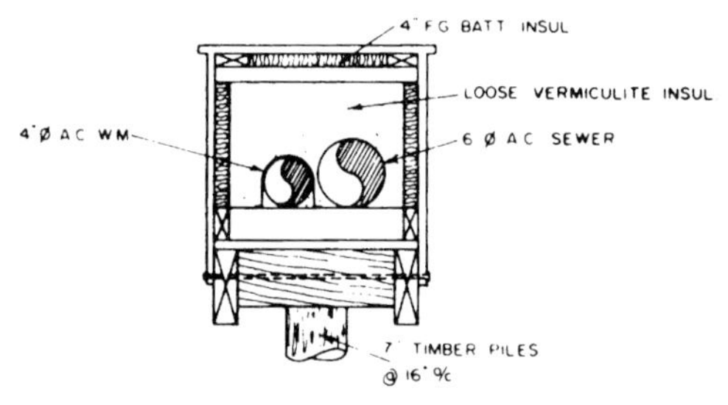

Figure 6. *Diagram showing typical utilidors, such as those used in Inuvik (Diagram from EPEC Consulting Western, Ltd., 1981).*

*   *   *   *

During the late 1950s and the 1960s, many families from fishing, trapping and hunting camps migrated to Coppermine, where the government provided houses for them. A reputable municipal engineering firm in Edmonton, planned the community. Unfortunately the planner made a few mistakes. Although he had visited the site to prepare his plan, he planned east-west streets across the rocky area to the east of the settlement. Since he was unaware of the rocky area, he could not have considered the elevation contours in making the plans.

Figure 7. *Caribou antlers. Photo by Jack Grainge.*

Figure 8. *Don Ayalik and skinned foxes. Photo by Jack Grainge.*

Dr. Gordon Butler, Director of Northern Health Services, stationed in Edmonton, asked me for my comments. I said the plan was impossible, because the east end of the streets ran into impenetrable rocks. I added that streets should run obliquely, northwest down the slope at a one percent grade. This would allow ditches, and future subsurface water mains and sewers to run down each road, draining readily and with a minimum number of street crossings. At the lower end of the streets, the sewers would be connected together so the sewage would flow by gravity to the sea.

A senior engineer in Ottawa wrote a letter to Dr. Butler justifying the original plan. He did not think that in view of the community being small and the winters being cold, piped water and sewer systems would ever be practical. I based my opinion on the fact that Dawson, Yukon Territory, with colder weather, had subsurface, piped water and sewer systems.

About two years later, an engineer of the Department of Indian and Northern Affairs and I were in a plane that landed in Kugluktuk for a gas refill. We had no time to look around because we had to reach Holman before it was too dark for landing. The community manager met us and I asked him why the streets were running north and south directly down the hill, when the plans called for them to run east and west. He laughed, "Those plans were impossible so I junked them. I just ran the streets in what I thought was the best way." So much for my suggestion for Kugluktuk to be eventually served by practical systems of piped water mains and sewers. It shows how difficult it is to make plans in Edmonton, Ottawa and Yellowknife, three widely spaced cities, for an installation in a remote northern community.

Figure 9. *Aerial view of Coppermine, showing the poor plan by the local administrator. Photo by Jack Grainge.*

* * * *

In January 1970 I returned to the settlement with a DI&NA party of engineers. In sixteen years the population had grown from 20 to 550. The nursing station had been replaced by one with four patient beds and rooms for four nurses. Also newly built was an additional seven-classroom school, a transient center, two RCMP houses, a Northern Canada Power Commission, diesel-electric, power plant, 105 low-rental houses, and a community bathing and laundry house. A local entrepreneur was building a motel with separate, individual units. He was also the agent for Northwest Aviation, and his wife was the postmistress. There was an airstrip south of the hill behind the settlement.

The Department of Indian and Northern Affairs had constructed an all-season water pipeline to the school, nursing station and the power house. Piping water to buildings was better than delivering water to household tanks. Most water tanks were subject to contamination from dirt falling into them. Some leaked and some overflowed. Most were difficult to flush clean. At that time wastewater, toilet sewage in plastic bags and garbage were hauled to a remote point alongside the road to the airstrip.

Kugluktuk will continue to grow. Perhaps some people will build houses on the south slope of the hill overlooking the airstrip. If so, the builders should arrange the buildings so that eventually they can be economically serviced by shallow water mains and sewers. It would be a sunny site for a residential subdivision.

References

Clark, L. 1996. Personal communication.
Grainge. J. 1970. *Report on Sewage Disposal, Coppermine.* Edmonton: Associated Engineering Services Ltd.
Lord, C. 1996. Personal communication
Lord, D. 1996. Personal communication.
Merrill, C. 1996. Personal communication.

RESOLUTE BAY

The community of Resolute Bay is situated on the south coast of Cornwallis Island, on the northeast shore of Resolute Bay. It is approximately 1,555 km from both main NWT air bases at Yellowknife and Iqaluit (previously known as Frobisher Bay). It was named after HMS Resolute, one of the British ships in the search for Sir John Franklin.

Artifacts show that about a thousand years ago, the island had been a hunting and fishing area. In 1947 USA and Canadian meteorological services constructed a weather station and airstrip there. In 1953, to take advantage of the game in the region, Inuit were moved there from Port Harrison and Pond Inlet. Hunting and fishing were productive, so in 1955 relatives came to join them.

Figure 10. *Resolute Bay (power plant in foreground). Photo by Jack Grainge.*

When I arrived in July 1973, there were three communities. The Inuit lived on the point south of the Bay. At North Camp the airport runway ran beside the Department of Transport (DOT), weather and communication station, and a long building containing a large restaurant, recreation room and bedrooms for both DOT personnel and transients. Offices, warehouses and houses of air transportation companies were nearby. The RCAF Arctic Survival School, maintenance buildings and fuel oil tanks were located at South Camp, near the wharf on the west shore of Resolute Bay.

All the communities are situated on a mixture of gravel, sand and soil of former beaches, now risen well above the shore. The land slopes gently from the airport to the sea shores to the south and east, with several lakes in between.

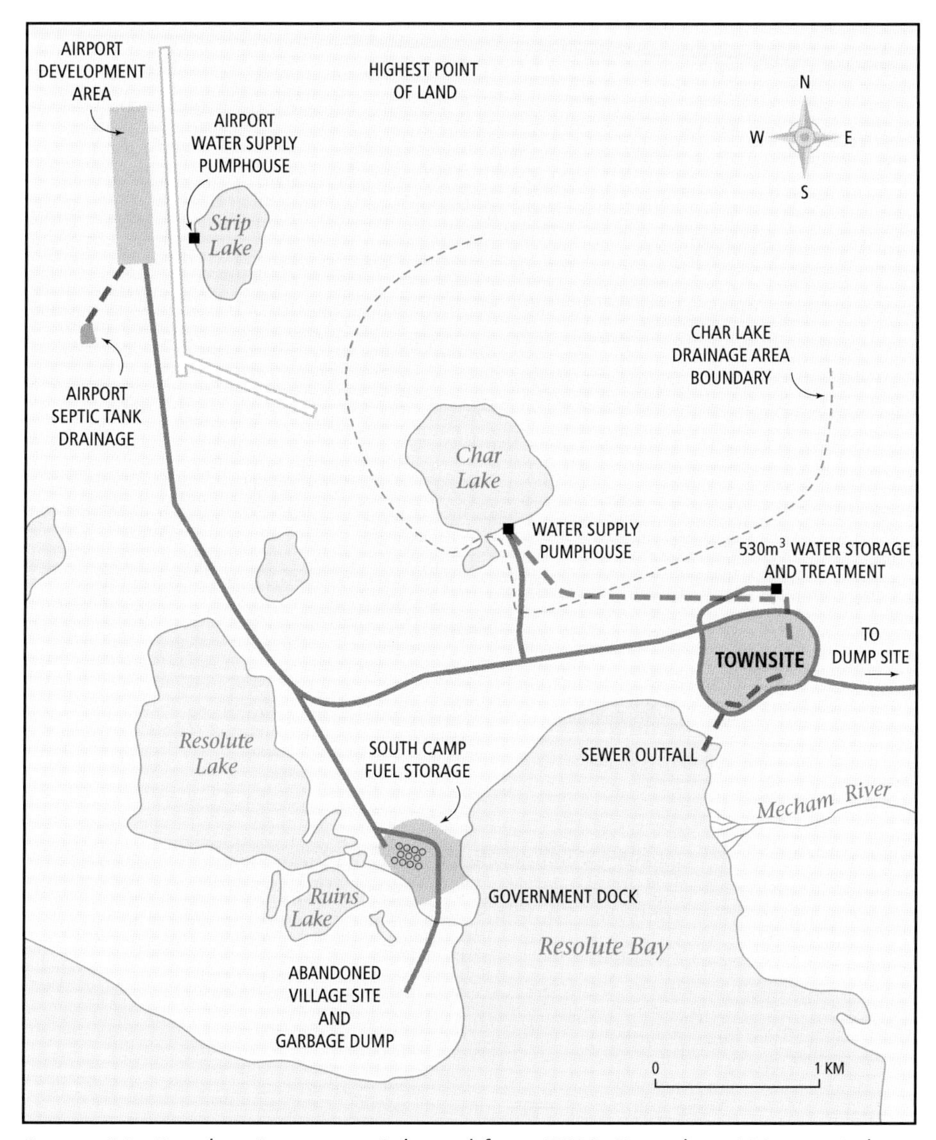

Figure 11. *Resolute Bay area. Adapted from EPEC Consulting Western Ltd. (1981), by Johnson Cartographics, Inc. Edmonton.*

I arrived on a scheduled, Pacific Western Airlines 737 airplane from Edmonton, stopping at Yellowknife and Cambridge Bay en route. A few days later I had finished my work and was awaiting a return flight. The skies were clear but intermittently clouded over. I happened to be in the DOT, communications office when I heard the Pacific Western Airlines plane directly overhead. The operator told me, "Don't go to the airport. The sky has covered over so the plane can't land." The airplane motors faded into the distance.

Fifteen minutes later the sky cleared. The operator began talking about other matters. I interrupted, "Aren't you going to advise the pilot that the sky is clear so that now he can land?"

"No. He wouldn't come back," he replied. "His passengers waste a day in the airplane, but he won't circle a few minutes waiting for the sky to clear."

The next day a laborer waiting to go away on vacation and I walked up a nearby hill behind the airport settlement. From there we looked at the low sun surrounded by soft blue sky and scattered fluffy white clouds. To the south, shimmering light reflected from the many lakes and the sea waves. The pale blue, main airport buildings brightened the grey gravel surroundings. Junked machinery, a wrecked plane and a huge pile of discarded tires littered the area to the right of the buildings. However it was too far away to be recognizable as junk. Huge, black, oil storage tanks by the sea dominated the background of the community. After twenty years I can still close my eyes and see that magnificent view.

While my companion and I were pointing out to each other the many points of beauty, a soft mist began to form. We hurried down the slope wondering whether we might get lost if a fog should descend. We experienced no difficulty.

* * * *

The airport community consisted of DOT communications officers, weather observers, airport construction and maintenance personnel, RCMP, a Department of National Health and Welfare nurse, personnel of charter air transportation companies and transients.

The airport building accommodated almost everyone in the community. It was aligned parallel to the main airstrip, which is north-south. It consisted of a long, wide hallway off which a huge dining hall and two long dormitory wings projected to the west. At the north end of the hallway was a large recreation hall with a canteen, pool tables and a library. Another dormitory projected northwest off the recreation hall. Two large garages and a warehouse were spaced around the north end of that long building. For guidance during blinding blizzards, hand rails ran between a doorway off the dormitory to the garages and the warehouse.

* * * *

The source of the DOT water supply was Strip Lake, east of the main airstrip. Dual pumps in a small, heated shack on the shore of Strip Lake operated alternately, drawing water through one of two, side-by-side, subsurface, galvanized iron pipes. The water flowed through one of two pipes in a shallow buried, insulated, wood utilidor. In winter a trickle of warm water from a water heater discharged back to the lake through the second pipeline. This warm water prevented the water in both pipes from freezing.

The hot and cold water pipes and hot-water pipes to the radiators ran in a wood utilidor along the long hallway floor beside the outside wall. The pipelines branched down the hallways of the dormitories. The floors of the main hallway sloped up and down over these branch lines.

Sewage discharged into a galvanized iron tank below the floor of each of the adjacent washrooms, located along the long hallway. Submersible pumps in the tanks discharged the effluent through a pipe, suspended slightly below the ceiling. Outside the building an overhead, insulated pipe discharged the sewage near a former garbage dump. No one objected to the appearance of the disposal site and there were no flies to spread diseases. This disposal site would not be acceptable farther south. The operator said, "The system works. Let's not fix it." I agreed.

The Inuit lived in a group of houses toward the point south of the wharf. They bucketed water from nearby Ruins Lake and discarded their garbage into the community dump farther down the slope. Later the people moved from this community to a site northeast of the Bay.

* * * *

In 1977 the local, Inuit Community Council wanted to know why the sewage from this newly constructed community was not treated in the sewage plant building. The parts of the treatment plant were lying on the beach. At the time I had retired, but before my retirement I had opposed the installation of this plant. The NWT Dept. of Local Government hired me to go there to explain the reason for the plant not being installed.

The outfall sewer ran to the building that now contained only a small comminutor to grind up the sewage. The effluent discharged to a sloping concrete pad at the shore of the bay. I told the Council that before discharging this relatively small amount of sewage into the bay, the authorities had changed their minds about what treatment was required. When they planned the sewage disposal system, they considered that all sewage must be treated before being discharged into the sea. However, as the local people well knew, the body wastes of the fish, seals and whales in the bay were many times greater than the sewage from this small community. I added that insects and water plants consume the sewage. The fish, seals and whales eat the bugs that eat the insects and water plants. They agreed with me that the sewage did not require treatment, and went on to discuss other matters.

Since my previous visit, construction of this new community was well on its way, with many houses already occupied. Ralph Erskine, a Scottish architect-community planner practicing in Sweden, had designed the residential subdivision. He had located the new subdivision on the south-facing slope rising from the north end of Resolute Bay. I believe that Erskine was hired because in my lectures and published papers, I had mentioned his name while speaking highly of community planning of northern towns in Scandinavia.

Erskine had aligned the streets so that most houses face south. Thus in winter the occupants may view what little sunlight there is, the noon twilight in midwinter and at other times the low sun. Spring not only seems to come more quickly — it does. The snow melts on that slope before it does in the surrounding countryside. Also the high ground behind the community provides some protection from the howling north winds. Thus the people enjoy a relatively warm microclimate.

Erskine designed the roads of the Resolute residential subdivision to run obliquely down the slope at a slight grade. Thus future, insulated, shallow buried, gravity sewers could serve long streets of houses. Buried water mains and sewers do not cut up the community and the school playgrounds as do above-ground utilidors.

The source of water was Char Lake, a large lake on the opposite side of the hill from the community. Water was pumped to a water treatment plant at the top of the ridge behind the settlement.

Once, while I was delivering a technical paper at the University of Montreal, I showed a slide of two apartment buildings in northern Sweden that Ralph Erskine had designed. While I was pointing out the several benefits of those residences in a northern climate, someone in the audience shouted, "He's here. He is sitting beside me." Erskine was a visiting lecturer, and I was pleased at long last to meet him.

References

Brown, T. 1997. Personal communication, Plumbing foreman during construction.

CAMBRIDGE BAY

Cambridge Bay is situated near the east end of the south coast of Victoria Island. Archeological investigations have shown that it has been a meeting place of the Inuit for centuries. In 1839 HBC Chief Factor Warren Dease and Thomas Simpson named the place after HRH Adolphus Frederick, Duke of Cambridge. In 1851, Dr. John Rae found caches indicating many Inuit visited the area, and the following year Captain Richard Collinson reported two hundred Inuit in the vicinity. In 1918-20, Roald Amundsen navigated the northwest passage, leaving one ship, Maud, sunken near the shore in the bay and visible from the shore. I was surprised that it is no larger than a medium-size yacht. The next year the HBC opened a trading post, and Catholic and Anglican priests established missions. Five years later the RCMP established a detachment.

In 1946 the RCAF began an Arctic survival training school for airmen. The Canadian army constructed a LORAN, radar, navigational aerial but soon abandoned it. While constructing a DEW (Distant Early Warning) Line station there, the Northern Construction Co. demolished the high steel derrick of the LORAN antenna.

At Christmas and Easter Inuit converged at Cambridge Bay for religious celebrations, to trade at the HBC and to visit with friends. Although fish, white fox and seals were plentiful in the area, few people remained there after the celebrations.

* * * *

In September 1956, I made an inspection tour of DEW Line stations, including the one at Cambridge Bay. I walked two kilometers to the Cambridge Bay community. Amiable Ken Berry, Technical Officer, Department of Public Works (DPW), was setting up camp for a crew to build a residential school. The center of the community would replace the existing surface cemetery. Bodies, some wrapped in canvas and some in wood boxes, were lying haphazardly on the

Figure 12. *Surface cemetery in what was to become the center of the community. Photo by Jack Grainge.*

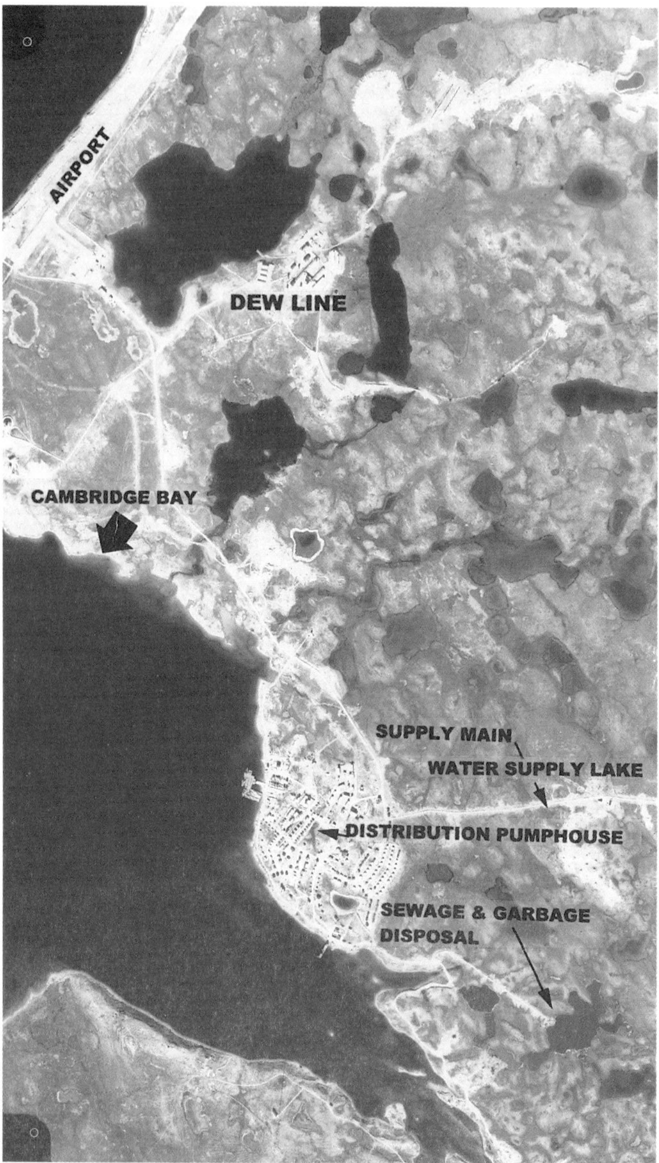

Figure 13. *Aerial photo of Cambridge Bay area, showing Distance Early Warning (DEW) line site. Department of Energy, Mines, and Resources, 1975.*

Precambrian rock. Later workmen moved the bodies to a sandy area where the ground could be thawed and the bodies buried.

The few outsiders in the community lived in buildings spaced widely around the planned community. The wood-frame HBC store, house and warehouse were near the original dock, at the east end of the present community. The HBC manager and his wife lived in a house two hundred meters northwest of the post. The newly built Anglican rectory was on high ground to the northeast of the school, and Father LeMer's stone church was near the shore southeast of the school.

The following May, I returned to Cambridge Bay on a Department of Transport (DOT), DC3 plane. Pilots

Johnny Sapphire and Tom Prescott were checking the operations of the DEW Line airplane controllers. They had invited me to come along. On the way the pilot shut off one of the two engines in order to check the other one. Louis Raymond, the agile, always-joking, airplane mechanic told me that the same thing had happened on a previous flight and one passenger had become nervous. He asked Raymond, "What should we do?" Raymond answered, "Repeat after me, Our Father, Who art in –."

In Cambridge Bay I visited three buildings on the opposite shore of the Bay. Corporal Jack King of the RCAF Arctic Survival School with headquarters in Edmonton, was cleaning up after the last class for the year had gone.

A few dozen snow houses and a few emergency snow shelters built for grounded airmen were proof that they would know how to stay warm. An emergency shelter is a row of A-frames, each A-frame made from two snow-blocks. In case of an emergency landing during a blizzard, a person could build one of these in an hour or so. King said that a man wearing winter clothing could crawl in feet first, and be warm. During a blizzard, he would have to wake up from time to time to ensure that the opening had not become blocked by hard-packed snow. He would not be able to dig out through hard, drifted snow. Also I think he would need added insulation under him.

Figure 14. *Rev. Leslie Corness and RCAF personnel. Photo by Jack Grainge.*

Father LeMer was living in a stone church-residence. He told me he had followed the example of Father Henry, who, in the mid 1930s, built a small stone church at Pelly Bay. Father LeMer spent two years gathering stones for the construction. The walls, made of stones and clay, were eighty centimeters thick. He built interior wood walls and partitions. He wired the buildings for electricity. He was indeed a multi-talented craftsman.

Three Inuit families were farther east. They lived in small, low ceiling huts, which these clever, untaught carpenters had made with plywood from packing cases. They obtained many kinds of building materials at the DEW Line trash heap. Huts though they were, they were better than snow houses. In the huts they could use oil-burning lamps and tabletop stoves. These were better than seal-fat lamps used for both lighting and cooking in snow houses. In cold weather they banked their huts with snow to make them draft-free. The Inuit wore warm clothing when indoors.

Canon Peter Emerson and his wife and baby lived in a new, wood frame house. I asked him why on a cold day he had taken off his front storm window. "Its ironical." he replied, "our living room is unbearably hot and the other rooms are cold." We agreed that some houses in the North should have fans and ducts to blow when needed, warm air from south rooms to north rooms.

Figure 15. *Fr. LeMer and his church. Photo by Jack Grainge.*

* * * *

The people hauled water from a small lake a half kilometer northeast of the proposed school. The planned water supply lake was a large lake two kilometers north. The water in the large lake was slightly colored and soft.

Eager to know how thick ice on a lake in the high Arctic could become, I borrowed a needle bar and chipped a hole through the ice on the large lake. On a bare, wind-swept part of the lake, the ice was 2.7 meters thick. Wow!

People used honey buckets. They threw waste water near the houses and hauled toilet wastes to inconspicuous low spots. They burned garbage in empty oil drums, and dumped the ashes nearby.

In the summer of 1966, Norm Lawrence, President, Associated Engineering Services Ltd., was asked to make recommendations for the water and sewerage systems to serve a proposed residential school. He told me he recommended piping water from the large lake, later named Water Supply Lake, and piping sewage to the sea.

The Department of Indian and Northern Affairs (DI&NA) contracted Fred Ross to haul water and sewage for the community. He had previously been doing that work privately. He had also been renting beds in a half tent and providing meals for his tenants. Ken Hawkins, Chief Engineer for DI&NA in Fort Smith, told me that they made Ross a generous contract upon his promise to improve his equipment and build a substantial hotel.

When I was there in January 1970, the population of the town had grown to 550, of which 450 were Inuit. Many Inuit had moved there from outlying hunting and fishing camps, and were living in wood-frame houses. The operator for Fred Ross and Associates was hauling water from Module Lake, bordering the DEW Line building. The water was extremely hard but clear and otherwise satisfactory. The water in Grienier Creek was not so hard and that in Water Supply Lake was soft. The main reason for hauling water from Module Lake was its sound access road for Ross's 9000-liter tank truck.

I recommended to the driver that he take better care to prevent the hose nozzle from dragging on the ground. He should roll the hose on a reel in the heated compartment on his truck. Also, the outlet tap on his sewage truck should be repaired.

By 1970 Cambridge Bay had grown with the addition of a DI&NA administration office, Northward Aviation Co. base, DOT airport building, eight-bed nursing station, four-man RCMP post, Canadian Telecommunications office, NCPC power plant, two schools, NWT transient center, Pentecostal Mission, handicraft center, eight-unit apartment building leased by the NWT from Solar Construction Co., two, six-person dormitories for school children, and eighty houses for Inuit.

Figure 16. *Sewage haulage tank being connected to a sewage tank in a house. Tank outlet dripping on the road. Photo by Jack Grainge*

Hauled water and sewage services for this large population resulted in difficult sanitation problems. The two school dormitories were typical. Each contained three double rooms for children and a room for a house mother and her husband. Each home contained a pressure water system drawing water from a 3500-liter, galvanized steel reservoir. The water from the lakes contained sediment that settled to the bottom of the tank. The outlet pipe was slightly above the bottom of the tank, so the sediment from each filling accumulated and the tanks could not be hosed clean. This was a fault common to all systems of this type in the North.

I recommended that water reservoirs should have sloping bottoms with drain plugs lower than the bottoms. Existing flat-bottom tanks should be tilted to one corner, with drains fitted on the undersides of the low points. The interiors of such tanks could then be hosed down. To prevent dust from getting into the water, the covers should have turned-down lips fitting around the top edges of the tanks.

In most cases sewage-holding tanks were headaches. The one in the heated crawl space below the nursing station overflowed occasionally. Consequently the wood floor of the crawl space was rotting. Buried tanks beside the houses, such as the one at the transient center, must be insulated and heated at great expense. The one at the transient center overflowed, soaking the crawl-space insulation so that the sewage froze.

The Inuit houses contained 710-liter water tanks with outlet spouts, the tanks resting on stools. This was an improvement on dipping from a barrel.

The people threw wash water wastes outside and workmen hauled away their plastic toilet bags. The people placed these bags outside, where often they froze to the ground. Then, when workmen lifted them, the bags split. The only time freezing to the ground was not common was for about five summer weeks and when the snow drifts under them were deep enough to prevent them from melting all the snow under them. Fred Ross's company hauled the toilet bags and garbage to a dump site.

People in government-supplied, wood-frame houses were not as careful in preventing cold drafts as they were years earlier. They used to be to avoid cold drafts. When I was in Cambridge Bay in 1957, the people built ice-block porches around their doors to the outside.

However, finding faults is easy. Designing systems and buildings, explaining to employees how to construct them, and trying to supervise their construction from distant offices, is difficult.

References

Berry, K. 1996. Personal communication.

Grainge, J. 1970. Two reports on Sanitation by Jack Grainge, Region Engineer, Public Health Engineering Division, DNH&W, based on a study made January 12, 13 and 17, 1970. Reports were in the files of Ev Carefoot, formerly an engineer in Associated Engineering Services Ltd.

Lawrence, N. 1996. Personal communication.

GJOA HAVEN

Gjoa Haven is situated on the southeast coast of King William Island on the shore of Gjoa Haven Cove. The land, mainly sand with some stones, rises gradually from the shore. Soil patterns show that the land is a series of raised, former beaches. Now, long after the heavy glaciers have receded, the ground continues to rise.

During his 1903–1906 navigation of the Northwest Passage, Roald Amundsen spent two winters there. He found it a protected harbor for wintering his ship, Gjoa. Later Inuit from Chantrey Inlet and Sherman Inlet, both on the mainland to the south, moved there.

In 1923 the HBC opened a trading post near Douglas Bay and in 1927 they moved it to Gjoa Haven. Also, before going out of business during the 1930s, the CanAlaska Trading Co. had a post there. After they left, a Catholic mission occupied their abandoned buildings.

Figure 17. *Leaving Gjoa Haven, 1960. Photo by Ev Carefoot.*

During August of 1960, Curt Merrill and Ken Hawkins invited me to accompany them on a flight to three settlements. Pat Carry was the pilot. A prominent sign in the plane read, "FASTEN SEAT BELTS." We complied but discovered that our seats were not fastened to the floor. I remarked, "I think we're jettisonable freight."

When we landed in Gjoa Haven, Ken Hawkins said to no one in particular, but obviously intended for my ears, "Don't anyone make a recommendation to move any of the communities we will be seeing." I suppose the mandarins in Ottawa were still numb from the astronomical cost of creating Inuvik after I had recommended that Aklavik be moved.

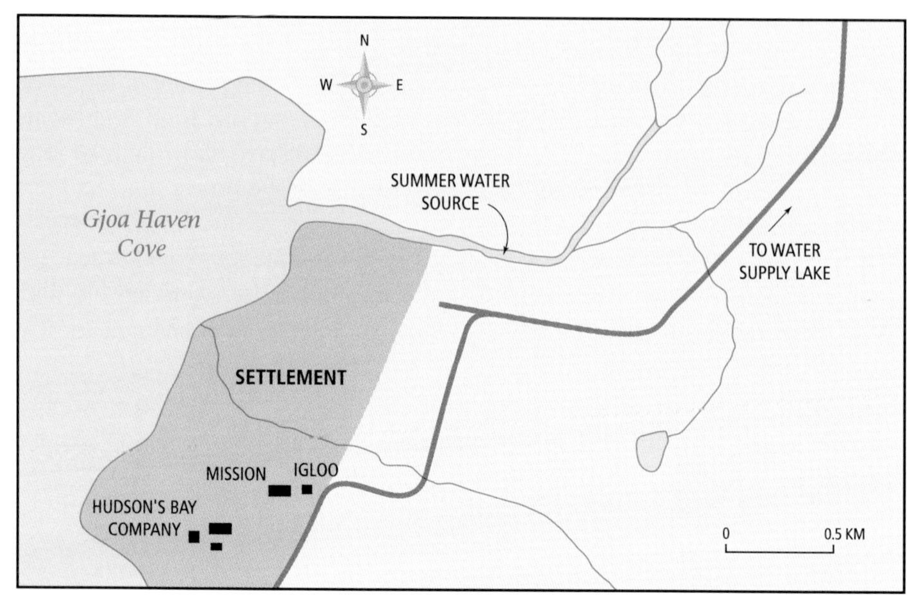

Figure 18. *Map of Gjoa Haven settlement. Adapted from EPEC Consulting Western Ltd. (1981) by Johnson Cartographics Inc., Edmonton.*

We spent a few hours in Gjoa Haven. I spoke to the manager of the HBC store, who was the son of a former manager of an HBC post and an Inuit. I said, "Hello, George Porter. I met you once before." He surprised me with his recall. He replied, "DEW Line airport, Gladman Point, May 23, 1957." I remembered his name because he was well known in the North. I was dumfounded by him remembering me and even the date of our meeting.

Figure 19. *Ken Hawkins beside George Porter and another Gjoa Haven resident. Photo by Jack Grainge.*

Porter's store was crowded with people, but

Figure 20. *Dogs, people, and sledges. Photo by Jack Grainge.*

Figure 21. *Father Henry and residents of Gjoa Haven. Photo by Jack Grainge.*

he immediately offered to show me something. When I questioned his leaving the store, he replied that while he was away, no one would touch anything.

He took me to the shore about two hundred meters from the HBC dock. He pointed to a place ten meters off shore. He said that a few days earlier, some kind of a machine had partially emerged from the sea. He imitated the noise of its motor. He thought that it might have been either a Russian or an American submarine. He wired the RCMP. Two days later two soldiers came to see him about the matter. They listened to what he said but made no comment. He heard nothing more.

In May 1967, I visited Gjoa Haven together with Dr. Davies and Nurse Kay Keith. Since my previous visit, the settlement had grown considerably. I looked around but remember little of the visit. Father Henry showed me a large igloo that he had made from his home-made concrete blocks. It was large enough for all of the people to assemble for an evening of dancing to their chant-like songs.

In previous years the various families would come together at Christmas and Easter to celebrate and to sell their furs.

A large hunting party, with sledges hauled by two teams of dogs in fan harness formation, returned from the continent, sixty kilometers distant. They had caribou carcasses, furs, and fish.

The people were friendly and we communicated with gestures. Sociologist Jamie Bond told me that the trip they had taken was dangerous. The people carried water from a nearby small stream.

Figure 22. *Returning home from a hunt on mainland. Photo by Jack Grainge.*

Figure 23. *Lady in front of hut, heaped with snow. Photo by Jack Grainge.*

We did not stay long, but the pilot said it was too long. He was apprehensive, because we had a long way to fly before sunset. Our next stop was Spence Bay. I sat in front beside the pilot. As we approached, after the sun behind us had dipped below the horizon, he and I were looking back and forth across the darkening landscape. He worried that we might miss the place. However he found it, and landed the plane on the water without difficulty.

TALOYOAK (SPENCE BAY)

The hamlet of Taloyoak, formerly known as Spence Bay, is situated on the west coast of Boothia Peninsula. It lies beside a beautiful lake and extends around Spence Bay Inlet. Being on the Pacific side of that peninsula, the tides are only a few centimeters high. Large rocks, interspersed with patches of sand, clay and gravel, cover the area.

In 1904 Roald Amundsen explored the area on his east-to-west voyage through the Northwest Passage. Captain John Ross wintered in the area during his 1829 to 1833 explorations. Felix Booth, a wealthy gin distiller, financed the expedition. Boothia Peninsula was named after him.

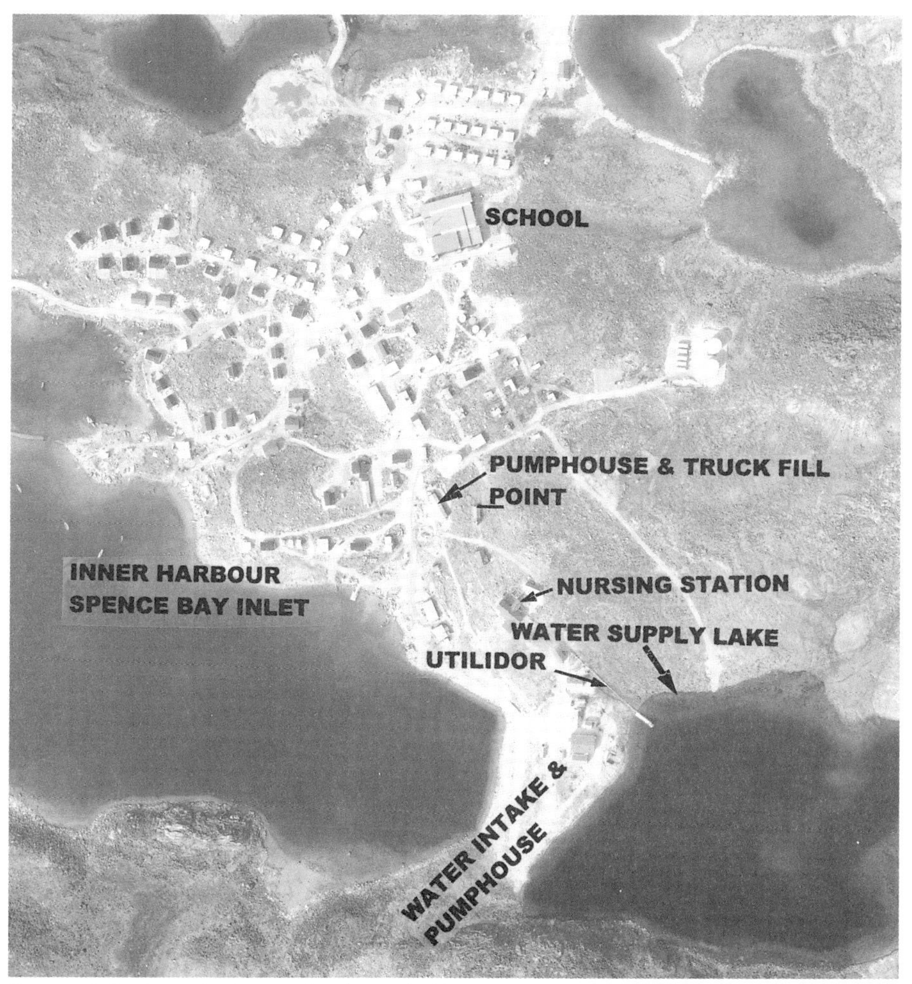

Figure 24. *Aerial photo of Taloyoak (Spence Bay). Department of Energy, Mines, and Resources Ltd., 1977.*

In 1934 the HBC moved fifty-two Inuit and their one hundred and nine dogs in their ship, Nascopie, from Cape Dorset and Pond Inlet to Dundas Harbor, on the southeast shore of Devon Island. The site did not prove to be satisfactory so the following year the HBC moved them to Crocker Bay, seventy-seven kilometers farther west. In 1936 they moved them again, this time to Arctic Bay. A year later they took them to Fort Ross, a new post on the south coast of Somerset Island, where they remained until 1947. Fort Ross was difficult to access, so the HBC closed the post and moved the people to Taloyoak.

They settled on the east side of the end of Spence Bay Inlet, while the Netsilik Inuit, the school and administrator occupied the west side. The HBC post was on the neck of land between a lake and the tip of the inlet. The two groups spoke different dialects and therefore had few contacts together. The Netsiliks referred to the newcomers as Dorsets. The situation changed when the children

Figure 25. *Hudson Bay dock/buildings. Photo by Jack Grainge.*

Figure 26. *Snow house, May 1963. Photo by Jack Grainge.*

Figure 27. *Hut buried by snow for insulation, May 1963. Photo by Jack Grainge.*

attended school together and the younger people of both groups began speaking English.

That was the first time I realized that there were five Inuit dialects. I later experienced that when John Shaw and I had a scout troop in the tubercular ward of the Charles Camsell Hospital in Edmonton. I was eager to give a translator's badge to two Inuit boys. I gave one a message in English and asked him to repeat the message to the other boy in Inuktitut, but not in English. They failed the test because they did not speak the same dialect. In the hospital all of our Dene and Inuit boys spoke English to one another.

Throughout the travels of the Dorsets, Ernie Lyall, a HBC man from North West River, Labrador, stayed with them and helped them. He married Annie, a Dorset, and they raised several children. After his retirement, he and his wife lived with their daughter and family in Leduc, Alberta. I was never able to contact him because I did not know his daughter's surname. Later

Figure 28. *Dogs and sledges, May 1967. Photo by Jack Grainge.*

Figure 29. *Snow house with nursing station in background on high land. Photo by Jack Grainge.*

he returned to the North. After retirement I did some work in Labrador, where I met his relatives. They had lost touch with him.

In May 1963 Nursing Supervisor Kay Keith and I, while visiting several stations, arrived in Taloyoak in a small plane. The next day, Smokey Hornby arrived to fly me to Hay River. He said that that town was flooding, and I had to get there quickly. The next time I was in Ottawa I received a dressing down for not knowing beforehand that a flood in Hay River had been imminent. I could only unconvincingly reply that when I had left Edmonton a few days earlier, there had been no mention of that danger in Edmonton newspapers.

* * * *

While in Taloyoak I noticed that one family was living on the shore at the upper end of the water supply lake. Their sewage had nowhere to go but into that lake or to the land and be washed into the lake. I made a report that resulted in the family moving to the rest of the community.

Figure 30. *Inuk special constable constructing sledge for RCMP. Photo by Jack Grainge.*

There was a large nursing station on a high point overlooking several houses for government officials. The nursing station had a water reservoir, a water pump, plumbing, and a sewage holding tank, all in a large, high space below the main floor.

Figure 31. *Inuit in furs building a snow house. Photographer unknown.*

I suggested that they install a pump and pipeline from the water supply lake. The pump should be in an insulated and heated small house. There was no soil on the rock, so the pipeline to the nursing station, in insulated boxing containing an electric heating cable, should be laid on the ground surface. After each filling of the tank, the pump and pipeline should be drained and the heat turned off. The sewage tank was to overflow to drain among the rocks to the sea.

The next summer Jack Gemmel, the maintenance manager for nursing stations, began installing those facilities as well as building a garage. I suggested to the HBC manager that since the water pipeline passed close to his house, he should connect to it. I said that he could install an insulated water tank in his attic and enjoy running water. The manager started making the connections. Gemmel stopped that. He did not understand that in small communities, people helped one another in whatever way they could. I hope that after Gemmel left, the nurse reversed Gemmel's order.

After I was last there, the territorial engineers hauled gravel to fill the low spots in the settlement. I believe they installed surface running water and sewer systems to serve the school and teachers' residences.

Sometimes community nurses and others wrote to our office asking questions. To save time, we prepared individual sheets dealing with typical problems. We then forwarded these to inquirers along with short covering letters. For a centennial project in 1967, we assembled the sheets in a seventy page Sanitation Manual for Isolated Regions. We sent copies to Indian Reserves in Alberta and communities in the Northwest Territories. We were surprised by the demand, including hundreds of requests from hippies throughout USA and Canada.

In June 1963 David Tuktoo, an obliging, affable Inuit, became the DNH&W community health worker. The people in the community chose him, having been asked to recommend a young person whom they respected. At Cambridge Bay he took a three-month course about teaching sanitation. Ethyl Martens, a health educator in the Department of National Health and Welfare, Ottawa, taught the course. Mike Pich, a friendly, capable engineer in our office, helped her. I thought it would be a good way for Pich to become acquainted with sanitation problems for which he would later make suggestions for correction. The obliging Pich started the job a week after he became married.

Figure 32. *David Tuktoo beside garbage dump that he blasted out of solid rock, September 1963. Photo by Jack Grainge.*

In May 1967, Tuktoo was fluent in English, and did good work in helping his people adapt to the outsiders ways. The Anglican minister also appointed him a lay reader. Years later Tuktoo became Council Chairman.

Tuktoo invited me for an evening visit in his home, which he, with neither experience nor instruction in carpentry, had built. His wife and eight year old son were friendly and all smiles, but they could speak no English. His son demonstrated a Cossack dance for me. With arms folded in front of him he squatted with one leg extended forward. With a hop he reversed legs, squatting on the one that had been extended and extending the other. Back and forth he continued while I clapped in unison. Tuktoo said that was one of their traditional dances.

A year or two later Tuktoo and five other community health workers came to a staff conference in Edmonton. One evening they came to my house for dinner and to watch TV. They could not understand one another's dialects, so we all spoke English. We had a great time.

PELLY BAY

The community of Pelly Bay is situated at the northeast end of the Canadian mainland, off the Gulf of Boothia, on the east shore of Pelly Bay. Being on the Atlantic side of Boothia Peninsula, the tides are up to seven meters high. The community is situated on a flatland, surrounded by rocky Precambrian outcrops.

As it is a good sealing area, Inuit have lived there for centuries. In 1829 Sir John Ross wintered there. In 1935 Father Henry arrived. During the first winter he must have endured severe discomfort living under a temporary shelter over a shallow excavation and the low beginnings of the walls of a small, stone-wall church-residence. The following year he finished it with a wood-frame roof. A few years later he built a larger stone-wall church.

Figure 33. *View of Pelly Bay, 1960. Photo by Jack Grainge.*

Figure 34. *Father Henry and RCMP Officer in front of stone church, 1960. Photo by Jack Grainge.*

In August 1960 I accompanied Dr. Bill Davies and Nurse Kay Keith on their annual round of visits to central arctic settlements. When we arrived in Pelly Bay, almost everyone was in church, so I joined them.

Figure 35. *Duncan Pryde beside half snow house, May 1967. Photo by Jack Grainge.*

Father Franz Van der Velde said the mass. The parishioners, all Inuit, stood and sang in Latin. At least I recognized some Latin. Afterward Father invited me to the rectory. He had a well-stocked library, mainly comprised of books about the North.

I visited a few Inuit in their tents. We spoke using gestures. They may have been able to sing and chant in Latin, but no one seemed to speak English. They carried water from shallow, nearby lakes. Scavenging dogs and birds took care of the garbage and toilet wastes. There were no cans or bottles scattered about. We left after a few hours.

Figure 36. *New wood-framed church/assembly hall, May 1968. Photo by Jack Grainge.*

In May 1967 I returned in the company of Dr. Brian Brett and Nurse Kay Keith. The colorful Duncan Pryde, campaigning to be elected NWT councillor, hitched a ride with us. He won the election.

When Pryde was seventeen, a graduate of a school for orphans in Scotland, the HBC recruited him as clerk in their northern trading posts. A clever fellow, he learned to speak all five Inuit dialects, build snow houses, trap and hunt, run a dog team, find his way across snow-covered country with few land marks and write an interesting autobiography.

Pryde and I visited Father Van der Velde in his new, wood-frame, spacious church-assembly hall. The Father was an affable, informative host.

We visited several people in their snow houses—not called igloos by Canadian Inuit. Before building a snow house, an Inuk (singular form of Inuit) poked a long, thin probe into snow drifts to find one with the right pack for making snow blocks. Packed snow, consisting of air trapped within the tiny spaces within snow crystals, may be the best insulation material in the world. Using a snow knife, which has a blade, sometimes serrated, almost twice as long as a butcher knife, he cut twenty centimeter-thick, oblong snow blocks from what will become the walking area of the house.

Figure 37. *Pelly Bay Mission (1960), featuring the stone buildings started by Father Henry. Photo by Jack Grainge.*

He set the first row of snow blocks end to end in a circle, the tips shaved so that they slope slightly upward. When he completed the circle, the line of their tops was an even slope rising the width of a snow block. He cut the bottom of each block with a small lip below one end. He butted the opposite end against the previous block. He set the end with the lip with a light tap on top by the heel of his hand. That tap caused the block to adhere firmly to the block below it. The chain of blocks spiraled upward and increasingly inward, completing closure at the top with the key block.

He made the peak blocks thin. This allowed heat to escape. This escape of heat prevented the inside surface of the upper blocks from melting and later freezing. It was necessary to do so because ice is a poor insulator. When ceiling blocks became iced, the igloo became cold and had to be abandoned. Alternatively he allowed heat to escape by setting a long, open-ended can through the peak block. Sometimes he used a piece of the hollow femur bone of a large animal.

Often an Inuk reduced freezing of snow by lining the interior with canvas (in previous years by seal pelts). He held the canvas in place by twine that he poked through the snow wall. On the outside he anchored the twine with sticks. A snow house is indeed a most ingeniously designed, cold-weather dwelling.

Figure 38. *Woman sitting on bed inside snow house, 1963. Photo by Jack Grainge.*

Figure 39. *Woman in sealskin parka. Photo by Jack Grainge.*

In some communities where driftwood or wood salvaged from abandoned ships was available, Inuit built snowblock walls to a convenient height. Then they placed lengths of lumber across the top and covered them with pelts.

About half of the inside of a snow house was at the level of the original snow surface. This became the daytime seat for everyone and at night the family bed. It was a bit higher than standard beds. A small wing extended from each side of the bed into the walking area. These wings became counters

The entrance consisted of an external, crawl-size tunnel. Its base was at the level of the floor of the

Figure 40. *Elder woman inside snow house. Note dark canvass lining the walls and ceiling. Photo by Jack Grainge (1966).*

Figure 41. *Woman in half snowhouse. Photo by Jack Grainge.*

snow house rising to the surface at the outside end. The oval-shape roof of the entrance was made with snow blocks.

The tunnel was long enough for people to shed their outer garments and leave them there in the cold. If they brought them inside the snow house, the snow on them would melt. Then, when going outside, the moisture would freeze, and the garment would become brittle and cracked.

At night or when a wind causes draughts, a snow block was pulled in front of the opening.

Outsiders refer to this tunnel entrance as a heat trap. It reduced draughts. As in a freezer at a food store, the cold air remained in the lower part of the snow house and tunnel. The upper part of the snow house was warmer.

Dishes, food and the seal-oil lamp rested on one counter, with sewing materials and small items on the other one. The Inuk poked rods or sticks through the walls to support lattice shelves above the seal-oil lamp(s). Before going to bed the Inuit set clothes to dry on the lattice shelves.

The family slept on a blanket of polar bear pelts, fur side up. They slept under a blanket of pelts, preferably caribou. A hair seal pelt, fur side down under a pelt of another animal, fur side up, could substitute for a bear pelt. Bear pelts are too heavy for a cover.

The family slept side-by-side, with their heads toward the middle of the room. They did not wear underclothes because it would have caused them to sweat. In her book *Down North,* Mena Orford

Figure 42. *Man and boat, Pelly Bay. Photo by Jack Grainge.*

Figure 43. *Scraping pelts while on roof of new house provided by DINA. Photo by Jack Grainge.*

describes the discomfort of getting up in a snow house while wearing a wet night gown.

The wife slept on the outside near the lamp. She woke up at about hourly intervals to adjust the wick, which consisted of arctic cotton or arctic willow fluff. An Inuit adage: do not marry a woman who sleeps too soundly or you will burn with her (in a tent).

Figure 44. *Sewing a tent. Photo by Jack Grainge.*

When we were at Pelly Bay, some of the Inuit had moved into double-canvas tents. Dr. Otto Schaefer told me that in early days they stitched a layer of moss between the two canvasses.

The Inuit had chained their teams of dogs outside, each team in a row on a snow drift. They spaced the dogs far enough apart to prevent dog fights. When sleeping or during winds, dogs curl up with their noses under their furry tails. Blowing snow covers them. With snow both under and over them, they remain warm.

During the mid 1960s, many families received wood-frame cabins—small, but two or three times more spacious than snow houses. Later still some received substantial houses with bathrooms, bedrooms and kitchens, all containing modern conveniences, even refrigerators. Yes, you can sell refrigerators to Inuit. Eventually some houses contained plumbing.

References:

Orford, M. 1957. *Down North.* Toronto: McClelland & Stewart, Ltd.
Schaefer, O. 1968. Personal communication.

HOLMAN ISLAND

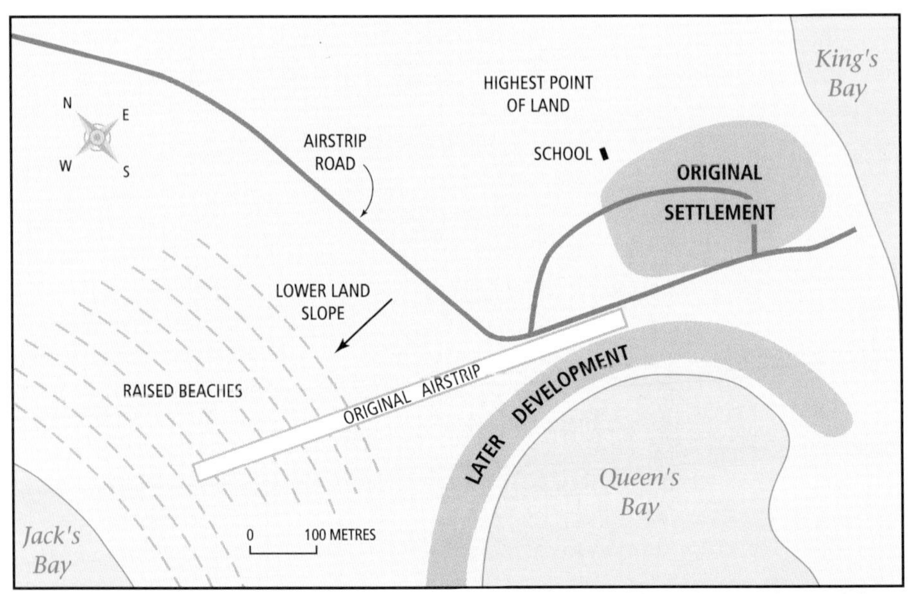

Figure 45. *Map of Holman Island, showing the raised beaches. Adapted from EPEC Consulting Western Ltd. (1981), by Johnson Cartographics, Inc. Edmonton.*

Holman Island, usually called simply "Holman", is situated on the southeast coast of the Diamond Jenness Peninsula of Victoria Island, between King's Bay and the north end of Queen's Bay. The soil is sand and gravel on raised, former beaches.

In 1911 Vilhjalmur Stefansson arrived nearby. He reported that there were two villages, each with about 150 Inuit. In 1939 Father Henri Tardi established a Catholic mission at Holman and Inuit began to visit there.

Figure 46. *View of Holman from the air. Photo by Jack Grainge.*

Figure 47. *Principal Born's daughter and Inuit friend. Photo by Jack Grainge.*

Figure 48. *Children in parkas, with fur hoods. Photo by Jack Grainge.*

The next year the HBC established a post there. An Anglican mission was built in 1962, and the first school in 1965.

In approximately 1968 I accompanied a Department of Indian and Northern Affairs (DI&NA) engineer and an administrator on a tour of communities. We landed on an airstrip running west through the middle of the community and continued across the series of raised beaches toward Jack's Bay. The pilot taxied right up to School Principal Born's doorway. There were no roads. The land slope was good and well drained, the ground was stable so roads were not necessary.

I believe the population was about two hundred. The buildings ran in parallel rows extending out from King's Bay. The school, three or four rooms, was located at the high end of the community.

A handicraft shop run by a bubbling-over, enthusiastic, young lady stocked Inuit prints and fur garments. I was lucky enough to buy a pair of wolverine mitts. The parkas and *mukluks* were attractive, but they would have been too warm in Alberta where I spent 97% of my time.

The school children's beautiful parkas, particularly the hoods, intrigued me. To admire and remember them, I took several photos of them. There seemed to be no Inuit men around the settlement. I presumed that they were away hunting and trapping.

For water the people melted ice blocks in water barrels in their homes. They cut the blocks from RCAF Lake, two kilometers northeast of the community, and brought them to the community by dog sled. In summer most people hauled water from the Ukpilluk River, 1.5 kilometers north. Others used ice they had stored in an icehouse.

Figure 49. *Night shot. Photographer, date not known.*

In September 1970, I accompanied Technologist Paul Mascho, Government of the NWT (GNWT), on a visit to Holman. We were on a plane with floats. Because it was getting dark we could not land on the bays near the community. We landed on RCAF Lake. There was an offshore wind so paddling was ineffective. I took off my trousers and towed the plane to shore. It was a cold, two-kilometer walk to a warm house in the community.

Many people had moved into the community. Houses had been built around the shores of Queen's Bay. If in future a decision is made to install piped water and sewer systems, the houses can be moved so that shallow, buried pipes can be laid on a gentle grade.

Several of the buildings, including the school and teachers' houses, had running-water systems. If my memory is correct, water for the reservoirs was trucked from RCAF Lake.

Sewage was hauled in a tank on a trailer from the holding tanks in the buildings. There were a number of low places to discharge the sewage. I suggested that a handy place would be any one of the dips in the series of raised former beaches overlooking Jack's Bay. Garbage was burned in abandoned oil barrels near the houses.

Those trips to Holman were interesting. I would have liked to return but the place was small and remote. I limited my visits to small, remote communities such as this, to occasions when I could share the cost of a plane with others. Unfortunately I had no such opportunity to return to Holman.

References

Berry, E. 1966. *Mr. Arctic (Vilhjalmur Stefansson).* New York: David McKay Co. Inc.

Carefoot, E. 1968. Personal communication.

Lawrence, N. 1968. Personal communication.

SACHS HARBOUR

Sachs Harbour is situated on rolling land on the southwest shore of Banks Island. The soil is clay containing many stones. It was named after the ship, *Mary Sachs* of the Canadian Arctic Expedition, that harbored there in 1913. Artifacts show that Thule culture Inuit once lived on Banks Island. During summers beginning in 1915, Inuit came there from the Mackenzie Delta to trap fox. Permanent settlement began in 1929 when three Inuit families sailed there in their own schooners.

In 1953 the RCMP established a post there, to be followed the next year by a DOT weather station. In 1958 Fred Carpenter, one of the hunters, constructed a store.

While I was in Aklavik in 1956, Fred Carpenter married Agnes Pfeffer. Rev. Gibson performed the ceremony at All Saints Anglican Church. The bride wore a white wedding gown and the groom a well-tailored suit. However, the church was one hundred meters away. It had rained the previous day so the streets were very muddy. Most of the distance was along streets that had been lightly graveled. Unfortunately the gravel, which had cost sixteen dollars per cubic meter, had sunk deep into the mud. The only vehicle in town was a pickup truck in which, I suppose, the front seat was muddy. The only way for Agnes to get to church without muddying her gown, was to ride standing up on the back of the pickup with Fred holding her steady.

Fred's first wife had died and like all Inuit, he needed a wife. A wife was a necessary part of a household. She was needed to raise the children, sew clothes, prepare pelts, cook and keep house. Agnes was younger than one of Fred's children.

I had the good luck to take a photo of Mary Carpenter, Fred's attractive daughter, wearing a parka. The double-fur halo, the inner one, dark wolverine, the outer one, tan colored long wolf hair, enhanced her natural beauty. Once she came to our house for dinner and proved to be an interesting conversationalist.

Figure 50. *Mary Carpenter. Photo by Jack Grainge.*

A few years later Fred Carpenter suffered a concussion while in a plane crash, while landing in a partial fog. He had disconnected his seat belt in order to point out the place to land. An hour later, Dr. Otto Schaefer and Noah Carpenter, a medical student and son of Fred, arrived in another plane. I visited Carpenter in the Charles Camsell Hospital, but his mind had not then recovered. I believe he did not recognize me.

In 1966, my friend Jules Cohen, a senior U.S. Public Health Service (USPHS) engineer based in Alaska, and I, together with Norm Lawrence of Associated Engineering Services Ltd. (AESL) and his son Bob, Danny Makale of Makale

Figure 51. *Photo showing bank erosion. Foreground: Julius Cohen (left) and Bob Lawrence. Photo by Jack Grainge.*

Community Planners, and a Department of Indian and Northern Affairs (DI&NA) engineer, flew to Sachs Harbour from Inuvik. We landed on the sea. The community consisted of a Catholic Mission, a DOT weather-and-communication station with ten regular and fifteen occasional employees, and about forty Inuit living in about fifteen small houses.

While we were there, fifty children from a residential school in Inuvik also arrived by plane. These children had spent the winter together in the residential school. They

Figure 52. *Father LeMer and Norm Lawrence outside of church-residence. Photo by Jack Grainge.*

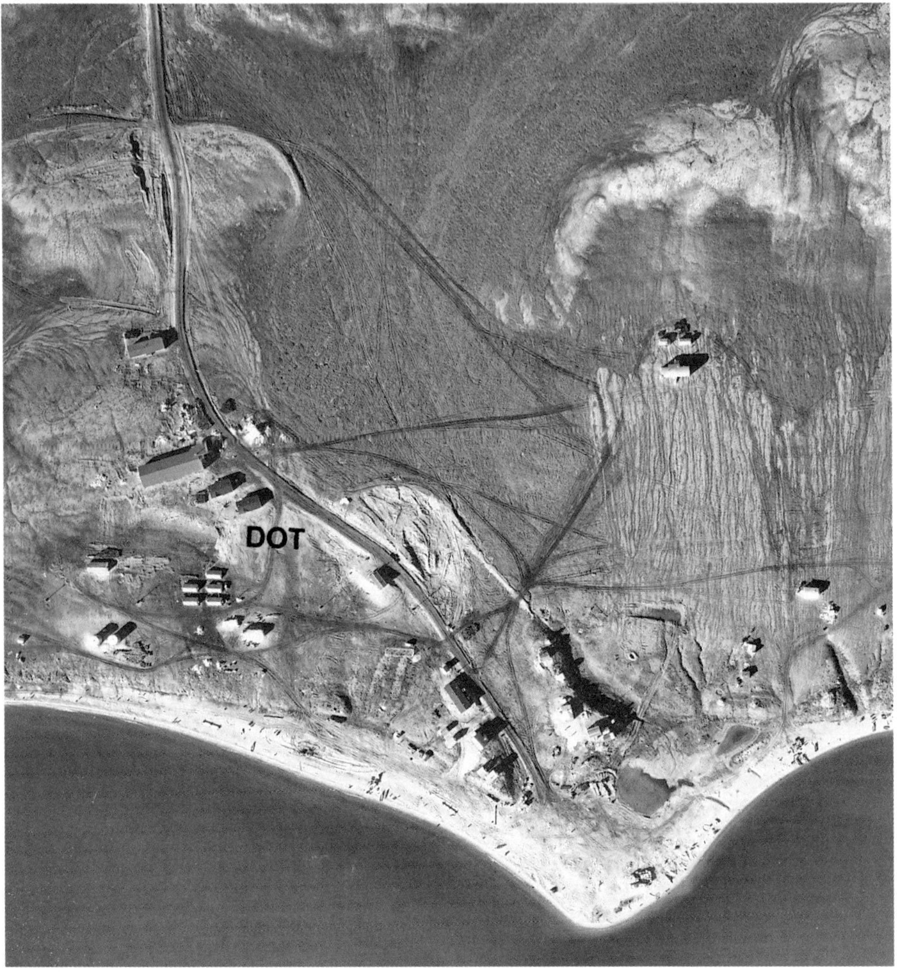

Figure 53. *Aerial photo of Sachs Harbour, showing DOT site. Department of Energy, Mines, and Resources, 1968.*

rushed home to greet their parents, but within an hour they were playing together outside.

Father LeMer did not take time out from painting his church. He said he had to take advantage of every one of the half dozen days a year of sunny summer weather.

We admired the industrious inhabitants, their well built, wood-frame houses and the cleanliness of their homes and the surroundings. Many of them were descendants of white whalers and Inuit mothers. The people were proud to be hard working and self reliant. David Nasogaluak, a successful hunter, told me

Figure 54. *Darren Nasogalak with arctic fox pelts. Photo by Jack Grainge.*

Figure 55. *Fred Carpenter with arctic fox pelts. Photo by Jack Grainge.*

about a welfare officer who had arrived the previous year. He called a meeting and told the people what he could do for them. However, they did not want welfare. They told him to leave. They considered assistance for old or sick people justifiable but not for healthy people. They were proud of their abilities to hunt and trap game, particularly the white and blue foxes that abounded on Banks Island, *their* island.

Carpenter's triple-pane windows surprised us. He said the extra pane made the house warmer. For many years the engineers in our office had been advocating triple-pane windows for northern houses. I discussed that fact in conversations with northern planners, in reports, in lectures and in semi-technical papers published in journals. Winter frost covers double-pane windows so that people cannot see out. Probably this confined feeling is partly the cause of many housewives becoming "bushed." However our advice was never taken in the NWT.

Triple pane windows have almost fifty percent greater insulation value than double-pane windows. Factories in Edmonton produce excellent triple-pane windows.

Lemmings abounded on the hill above the settlement. Apparently these rodents had been prolific under the winter blanket of snow. No wonder the arctic fox, their main predators, were so numerous on Banks Island.

During my northern travels, I have only seen lemmings one other time. In 1953 at Coppermine, I saw hundreds of them running about among patches of lichens. When I was there the following year, I saw none. Lawrence told me that in Sachs Harbour the following year, there were none that he could see.

* * * *

Figure 56. *Danny Mckale and Norm Lawrence at a meeting at the R.C. mission, June 1967. Photo by Jack Grainge.*

During summer and early winter, people carried water from a 2.7-meter deep lake behind the settlement. Also, they motor boated up the nearby Sachs River about thirteen kilometers to another lake. In midwinter the nearby lake froze to the bottom and the people melted snow in barrels in their kitchens. DOT personnel scooped up snow with a front-end loader. They dumped it through a trap door in a wall into a reservoir in their residence.

We were there for a day at the end of June. For more than two months the sun had been up for twenty-four hours per day. All the snow had melted, but a lake near the settlement remained frozen. Lawrence and his son bored a hole in the ice. It was ice all of the 2.7 meters to the bottom. Three weeks later they returned to find the lake had completely thawed. It is amazing what twenty-four hours of sunlight can do.

The one-day visit was all too short, but there was no accommodation for all of us. I had no opportunity to return to Sachs Harbour.

References:

Carefoot, E. 1968. Personal communication.
Lawrence, N. 1968. Personal communication.
Schaefer, O. 1968. Personal communication.

TUKTOYAKTUK

Tuktoyaktuk is situated on the southeast shore of Kugmallit Bay approximately 30 km east of East Channel of the Mackenzie River delta. It occupies a peninsula that forms the west shore of a bay known as Tuktoyaktuk Harbour. The peninsula rises two to fifteen meters above the sea.

Years ago an Inuit woman saw the reefs on the horizon, which she thought to be caribou. The Inuit named the place Tuktoyaktuk, meaning "Resembling Caribou." The community is commonly called "Tuk."

During the nineteenth century several hundred Inuit lived along the coast near Tuk, mainly near Kittigazuit and Atkinson Point. In his fascinating book, *I Nuligak,* interpreted by Father Maurice Metayer, Nuligak describes the life styles of the people of this area. Generally they constructed large houses using drift-wood and mud, each occupied by several families. On treks, wood was not available, so, they built snow houses. Their livelihood depended on whale hunting. From 1890 to 1910, many American whalers came there, which resulted in recurring epidemics of influenza. By 1920 influenza had virtually wiped out the Inuit in the western Canadian Arctic.

Figure 57. *Aerial view of Tuktoyaktuk. Photo by Jack Grainge.*

Inuvialuit from Alaska migrated east to the delta in two waves, the first in the 1920s and the second in the late 1940s, to obtain family allowances. Their dialect is called Nunatojmiut. Some of the Canadian descendants of those immigrants also receive dividends payable to Alaskans.

In 1937 the Catholic and Anglican missions and an HBC post became established in Tuk. The HBC began transshipping freight from river barges to seagoing ships in Tuktoyaktuk Harbour (Tuk Harbour). Also the HBC wintered their seagoing ships, Banksland and Hearne, in the shelter of Tuktoyaktuk Island (Tuk Island). In 1947 the Anglican Mission built a school, which the Department of Indian and Northern Affairs (DI&NA) took over the following year.

Figure 58. *DEW Line aerials with night sun behind them, 1956. Photo by Jack Grainge.*

In 1950 an RCMP post was established there, followed in 1955 by an Auxiliary DEW Line station, a nursing station, a DI&NA administration office and in 1962 by a fur garment shop and Mangilaluk School.

Figure 59. *Wall built to prevent bank erosion 1972. Photo by Jack Grainge.*

In 1955 the USA Army lent Northern Transportation Co. Ltd. (NTCL) six freighters with which to supply DEW Line stations through Bering Strait from Pacific ports. NTCL established an office at Tuktoyaktuk to control the distribution of the freight. Frank Broderick, my neighbor, was the general manager of the company. In 1960, Bruce Hunter succeeded him. Now deceased, Hunter lived near me in Edmonton.

* * * *

The soil at Tuk is silt, sand and gravel, laid down while the last glacier was receding. It overlays gravel river deposits from the previous glacial period. In recent centuries, delta silt from East Channel of the Makenzie River delta is being deposited along the south shore of Kugmallit Bay and along the shores

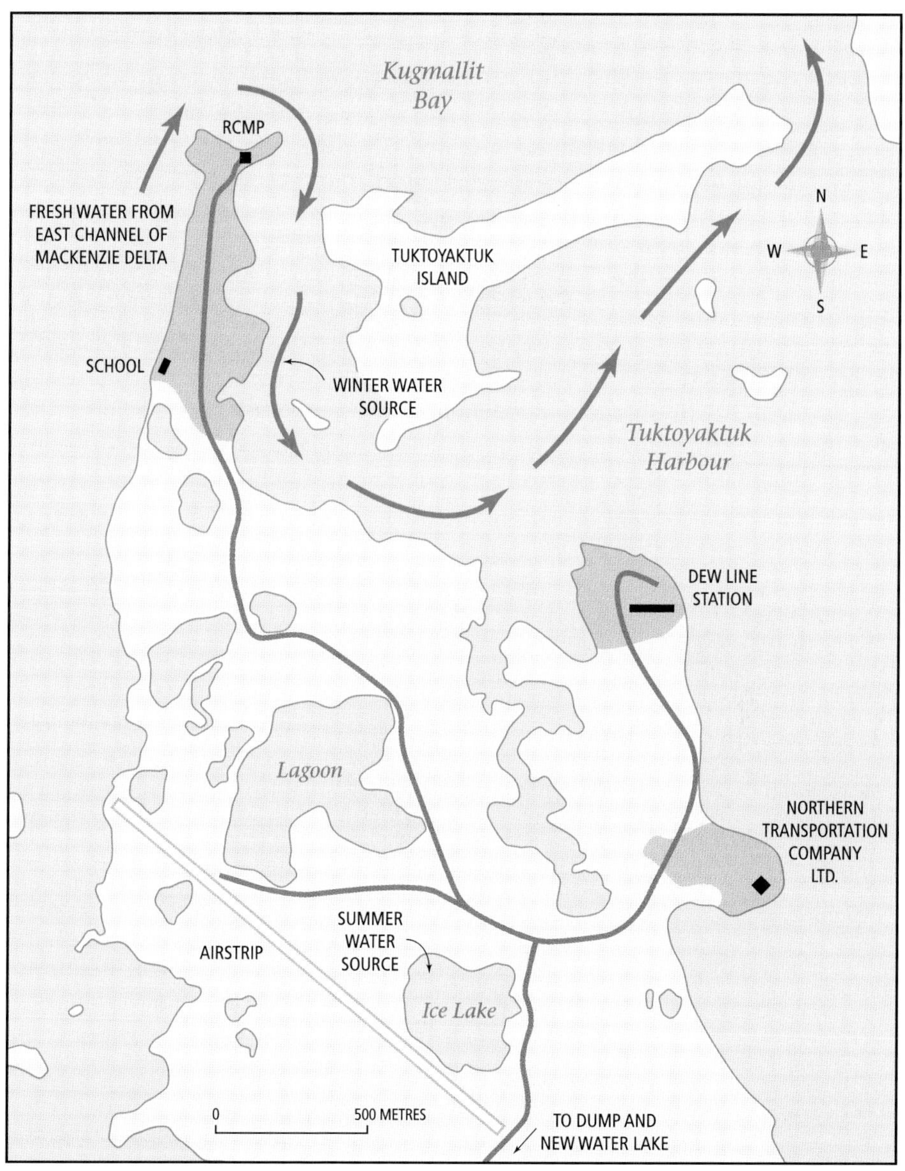

Figure 60. *Map of Tuktoyaktuk. Adapted from EPEC Consulting Western Ltd. (1981) by Johnson Cartographics Inc., Edmonton.*

of Tuk. Northern Construction Co., contractors for the DEW Line, constructed a gravel airport and roads using nearby gravel. After the removal of the gravel, underlying ice layers melted, resulting in the formation of lakes south of the air strip.

Since the first buildings were built, the west shore has been eroding, particularly along the peninsula near the RCMP buildings. In the 1960s, to stop erosion, the federal engineers built a wood-stave wall along the shore. The following year a storm washed away the ground from behind the wall. The next year the wall washed away.

I asked Charlie Walrath, an engineer with Department of Public Works (DPW), for an explanation. He said that usually erosion of beaches is caused by water currents running parallel to the shore, and that stub walls, projecting perpendicularly from the shore, will usually stop erosion.

* * * *

Any ground that remains frozen (temperature below water-freezing temperature) throughout the summers is permafrost. Depending on air temperatures, length of summers and insulation of surface soil and plant cover, the bottom of the permafrost varies from depths up to five hundred meters. Elsewhere pockets of permafrost forms in the ground below all-season rinks and frozen-food locker buildings.

Figure 61. *Inside the tunnel in the pingo at Tuk, 1956. Photo by Jack Grainge.*

During summers and winters, the interface between the frozen and unfrozen ground is alternately warmer and colder than the permafrost, causing moisture to migrate within the permafrost. This is the same effect as moisture migrating out of food in a freezer as the thermostat causes the air around the food to be alternately warmer and cooler than the food.

In places in both Tuk and Aklavik, people dug manholes through six

Figure 62. *Danny Makale and Doug Hargreave inside 9 m deep tunnel used for food storage. Photo by Jack Grainge.*

Figure 63. *The pingo at Tuktoyaktuk. Photo by Jack Grainge.*

meters of pure ice. Then they dug horizontal tunnels into the ice in all directions to form frozen-food storage spaces for the various families. Because there is little variation in the temperature at that depth, the food remains in perfect condition for long periods. It is like the well-preserved mammoths in arctic glaciers and the recently uncovered remains of a prehistoric man in an alpine glacier.

Yakutsk, Siberia, population 150,000, is located on a delta shore of the Lena River. A tourist attraction is the closed-and-insulated top of a hundred meter-deep well that a man had dug by hand. He encountered permafrost the whole way, but no water. Below a park in the same city, workmen had excavated ice to create two huge rooms to serve as laboratories for scientists studying the permafrost.

In the zone of discontinuous permafrost, from approximately Fort Norman to northern Alberta, permafrost occurs sporadically. For example, north-facing slopes receive less sunlight and are therefore cooler. Also thick beds of lichens are conducive to the formation of permafrost because they provide both insulation and dissipation of heat by both evaporation and transpiration.

In some areas permafrost uplifts the ground. The resulting mounds are called pingos. Most are small. One in Tuk is about nine meters high. The residents dug a tunnel into it, intending to make a curling rink. They dug the tunnel with a slight slope upward, making it easy to slide the ice out. Before they finished, warm air was rising upward in the tunnel, and thus melting the walls and ceiling. To prevent further melting, they insulated and boarded up the entrance.

Figure 64. *The pingo in Sondrestromfjord, 1960. Photo by Jack Grainge.*

The largest pingo in Canada, forty-five meters high, is three kilometers south of Tuk. Before its dome collapsed, no one knows how long ago, it was higher. In 1954, John Pihlainen of the National Research Council, Ottawa, bored eight-meter deep holes into it showing that the pingo was pure ice. Some Russian scientists for whom I was the guide in the Canadian North, told me that a pingo near the coastal mouth of a river in Siberia is more than 150 meters high.

In company with some engineers and scientists studying permafrost in Siberia, I saw a pingo about twenty-five meters high which grew and collapsed on a several-year, cyclical basis. They found upon investigation that there was a subsurface source of water, a spring in which the water did not rise to the surface. The water migrated close enough to the surface to feed the pingo. Eventually the pingo reached its maximum height.

The Russian scientists said that, during a dry spell, the source of water was insufficient to keep the pingo growing. Therefore it collapsed. In this case perhaps this happens. I think that as long as water seeps to the base of a pingo, it grows. Eventually its sides near the top reach critical steepness. On the south side, the sun thaws to below the surface of the lichens. A patch of lichens slough away exposing the ice below to the sun's rays.

I think the dome of the big pingo south of Tuk collapsed in this way. On the south-facing side of a high hill north of Sondrestromfjord, Greenland, I saw a meter-high pingo that was melting. The cover of earth, moss and lichens on its south side had melted and fallen away. Warm air was entering the opening, causing the pingo to melt. The bottom part of the core had already melted. A bridge of ice remained supporting the soil, moss and lichens on the top. The ice bridge was dripping. I guessed that complete collapse would occur within a week or two.

Terry Brookes, a territorial, municipal engineer, saw some ice mounds which may have formed in a remotely similar way. Two kilometers from Holman he noticed some snow-covered, meter-high pinnacles of ice sticking out from the ground. Perhaps water from subsurface sources fed them, and during the winter they grew.

* * * *

In 1956 Dr. Bill Davies and I flew to Tuk. The plane landed on a gravel airstrip, constructed along with the Dew Line station by Northern Construction Co. We stayed at the nursing station.

Ice Lake was the water-supply lake. It is near the southeast end of the airport. It was about three meters deep. During summer the water was slightly salty but drinkable. During winter, the ice froze from the surface downward. As the ice formed, it excluded some of the salt — the slower the freezing the greater the salt exclusion. The salt became concentrated in the water remaining below the ice. Consequently after a month or two the water became too salty for drinking.

In winter, ice forms on Kugmallit Bay, protecting the water from being riled up by winds, tides and waves. The water from the east channel of the Mackenzie River floats above the heavier salt water. It flows along the south shore of the Bay, around the tip of the peninsula, into Tuk Harbour. It makes an anticlockwise circuit around Tuk Island and out to sea.

At first the layer of fresh water is thin, but it gradually thickens. By Christmas the freshwater layer is thick enough for water to be bucketed from it. Then that becomes the winter water source. The most sanitary place to take water is after it has rounded the tip of the peninsula. The flowing water carries away the contamination introduced by the water buckets. Thus the water at the hole is always clean.

As a favor to the people, the driver of the DEW Line water truck delivered water to the reservoirs of the RCMP, nursing station and missions. The DEW Line operator chlorinated the water used by the station.

Al Smith, Superintendent of Buildings and Works, NTCL, asked me for suggestions for their water supply and sewage disposal. He planned to build a large NTCL base. For water I suggested that in summer he pipe water from Ice Lake. In the winter he should pump the salty water from below the ice into Tuk

Harbour, and then refill the lake with fresh water from the Bay. He constructed a buried, 76 mm aluminum pipe to Ice Lake for that purpose.

During a violent storm in the summer of either 1957 or 1958, salt water from the Bay washed into Ice Lake. The water in ice lake became too salty to be drinkable. For the rest of that summer most people boated across the bay to take water from Pikiolik Creek, where the water is soft and of good chemical quality.

Ken Hawkins, the chief engineer in Fort Smith, built a seven kilometer, lightly graveled, barely passable road to a lake south of Tuk. Later DEW Line workmen added more gravel to the road. The water in both lakes was salty, but the water from the distant lake was better. Smith, by following my earlier suggestions, was able to improve the water in Ice Lake.

Most of the government-owned buildings had buried sewage holding tanks beside them. There was much trouble with them freezing. I suggested they could use immersion electrical heaters such as farmers use to keep water in troughs from freezing. This suggestion was never followed, possibly because these heaters in septic tanks would be difficult to monitor and maintain.

The four-room school in Tuk had a sewage holding tank on the main floor. A pump, also on the main floor, automatically discharged sewage from the sinks and toilets into the holding tank. On one occasion the operator for sewage and water haulage failed to empty the holding tank before it became full. He had previously closed the pump-out opening. The sewage pump continued operating. With nowhere else to go, the sewage flowed up the stack. It spread about on the flat roof and leaked down the walls of two classrooms. Joe Harrison, the principal, had to crowd the students from those rooms into the other classrooms. Eventually the affected wallboards were removed and the framework treated with a lime solution.

By coincidence, Harrison later married my son's sister-in-law. At the time we did not recognize each other. However when, at a Christmas party, he recounted that experience in Tuk, I realized that we had met previously.

Sewage was discharged into the Bay, and the water current carried it away. Dilution was immense. In addition human bacteria have a short life in ocean water. For these reasons, I considered that the health hazard to fishermen was minuscule. At Jacobshavn, Greenland, workmen in a building built on piles, extending a few meters off shore, slit open toilet bags and dump the contents into the sea near fishing grounds. Nobody complains.

During summer, garbage was discharged south of the southwest end of the airplane runway. In winter the garbage was dumped on the ice in Kugmallit Bay. One year the winter dump was too close to the settlement. A wind carried the garbage on an island of ice into Tuk Harbour. The ice island drifted into Tuk Harbour, around Tuk Island and back to the sea.

I had always agreed with the people then in the North that ocean disposal of household garbage was satisfactory if it did not drift to a nearby shore and if it did not contain mercury, an extremely toxic metal. Those were the days before plastics in garbage was common. The discharge of plastics to the ocean should not be allowed because fish swallow bits of plastic and die as a result. When the Department of Environment was formed, I became an employee. I changed my recommendations to those of the DOE — no disposal of untreated sewage and garbage to the sea.

Dr. Gordon Butler, Director of Northern Health Service, DNH&W, continued to advocate disposal of garbage on the sea ice. He maintained that disposal on the sea ice got rid of more than half the year's refuse. He claimed land disposal sites in the Arctic could not be operated safely. Garbage on frozen ground contaminated the living environment. He said that garbage and sewage from New York, the largest quantity in the world, caused trouble on the nearby shallow ocean shelf. However that was no reason to forbid the discharge of the insignificant amounts of garbage from arctic communities to the seas.

In the early 1970s, the newly appointed NWT Water Board, consisting of representatives of several federal and territorial offices and two mines in Yellowknife, became the final authority. It ruled against sea disposal of garbage. That settled the issue for a few years.

In 1968 or 1969 Norm Lawrence of AESL wrote a report about Tuk. He recommended the preparation of a town plan that would accommodate increased growth. He suggested a large playground near the school. He said a nearby untended graveyard would make a good site if the graves were moved to a remote site. The graveyard remained where it was. Perhaps the planners had other plans for a playground.

* * * *

In the late 1950s, Department of DI&NA administrators decided that the Tuk Peninsula on which Tuk was situated, might erode away. To encourage settlement on the opposite shore of the Bay, they decided to build a wharf there. Ken Berry, a Department of Public Works (DPW) technologist who had constructed several wharfs for communities along the shores of the Mackenzie River, built this one. To prevent the ice gripping the piles and with rising tides lifting them, Berry set air bubblers near them. The bubblers caused water to circulate near the piles and thereby prevent ice from forming around them.

However people would not move to the proposed site. The DPW kept the bubblers going for two years and then quit. Within a year after turning off the bubblers, ice heaved the piles upward. In approximately 1971 Monte Stout, NTCL, bought the wharf for five hundred dollars, expecting the company might use it. Crown Assets, who were selling it, told Stout that they had valued it at a hundred thousand dollars.

* * * *

In approximately 1969 I boated across Tuk Harbour to the IOL base. They had made some repairs to the DPW wharf and were using it. They were hauling water from a nearby lake. The operator said they wanted to dispose of their sewage without polluting the water in Tuk Harbour. For public relations reasons, they wanted no criticism from any quarter. I suggested that they could create a sewage lagoon with no overflow by clearing some growth away from nearby ice-rich ground. The permafrost ice exposed would melt, forming a pond. However they did not have a bulldozer to strip the growth. I therefore suggested dynamiting a hole, to expose the underlying ice and allow it to melt. I heard later they followed my suggestion and it resulted in a successful sewage lagoon.

They had a garbage incinerator, approximately four meters high and four meters in diameter. The operator told me that feeding the garbage into it and raking the half-burned residue out was a big chore. He found that he could do a better job of burning with a bonfire. The site was more than a kilometer from the town. If the garbage was stored so that it would not blow away, and if they did not burn garbage on windy days, I said, "Why not?"

Over the years I have seen many garbage incinerators. Only those with gas or oil burners have been reasonably successful. All took a lot of work to operate.

On reading about the development of Tuk after my last visit about 1968, I marvel that there are so many new buildings. I am surprised, but of course they in-filled many lakes with stable soil.

References:

Carefoot, E. 1998. Personal communication.

Clement, M. 1998. Personal communication, former NTCL employee.

Hunter, B. 1998. Personal communication, former superintendent NTCL.

Lawrence, N. 1998. Personal communication.

Ripley, Klohn & Leonoff. 1968. *Community Granular Materials Inventory, Tuktoyaktuk.* Report filed with GNWT.

Schaefer, O. 1998. Personal communication

Stout, M. 1998. Personal communication, former Superintendent, Construction Engineering and Properties, NTCL

AKLAVIK

Aklavik is a fascinating Inuit name meaning "Habitat of Barren Land Grizzlies." I first heard the name from a friend who, in approximately 1939, had been a restaurant waiter on the HBC, paddle-wheel vessel, Distributor, sailing from Fort Smith to Aklavik. The vessel had three decks with the bridge in the top one. The bridge had to be high so that the captain could see ahead of the three or four barges that he would be pushing.

The settlement is slightly west of the middle of the Mackenzie River Delta. It is the only area in the Canadian Arctic where Inuit live below the tree line. In early times, small, isolated groups, traded and sometimes fought with Dene, who predominated in the southern part of the delta. Many species of animals and fish were abundant in the delta, and the muskrat pelts are the best in the world.

Figure 65. *Bush pilot and his Beechdraft plane, 1954. Photo by Curt Merrill.*

In the early 1900s, independent traders established posts in the delta. In 1912, the Hudson's Bay Co. built a trading post near log cabins of two Inuit families, Pokiak and Joe Greenland, across the channel from where Aklavik later became situated. In 1919 the Anglican Church established a mission, followed three years later by the western Arctic headquarters of the RCMP. In 1925 General Hugh A. Young moved the Royal Canadian Corps of Signals station from Herschel Island to Aklavik. The next year the Catholic mission was established, and the HBC moved their trading post there from across the channel. In 1929, the famous, bush pilot, "Punch" Dickins landed the first airmail plane in Aklavik.

In 1953 Stan Copp, my boss, sent me to Aklavik. He had visited the place in 1951. I hitched a ride on Magistrate (later Justice) Lawrence Phinney's *Wardair* charter flight from Yellowknife to Aklavik. Phinney was going there to hear an appeal from Dave Jones, an employee at the IOL bulk station. Lawyer John

Parker, Court Clerk Jack Gorrell and Pilot Rowdy Rutherford were also congenial companions.

Gorrell told me that on an earlier flight, he and Magistrate Phinney had sat behind two men who were plotting how they would push a certain mining stock so that its price would rise. Gorrell cocked his head so that he heard the details. The next morning he bought some shares. A few months later when I met him in Edmonton, he told me he made a pack of money on that stock.

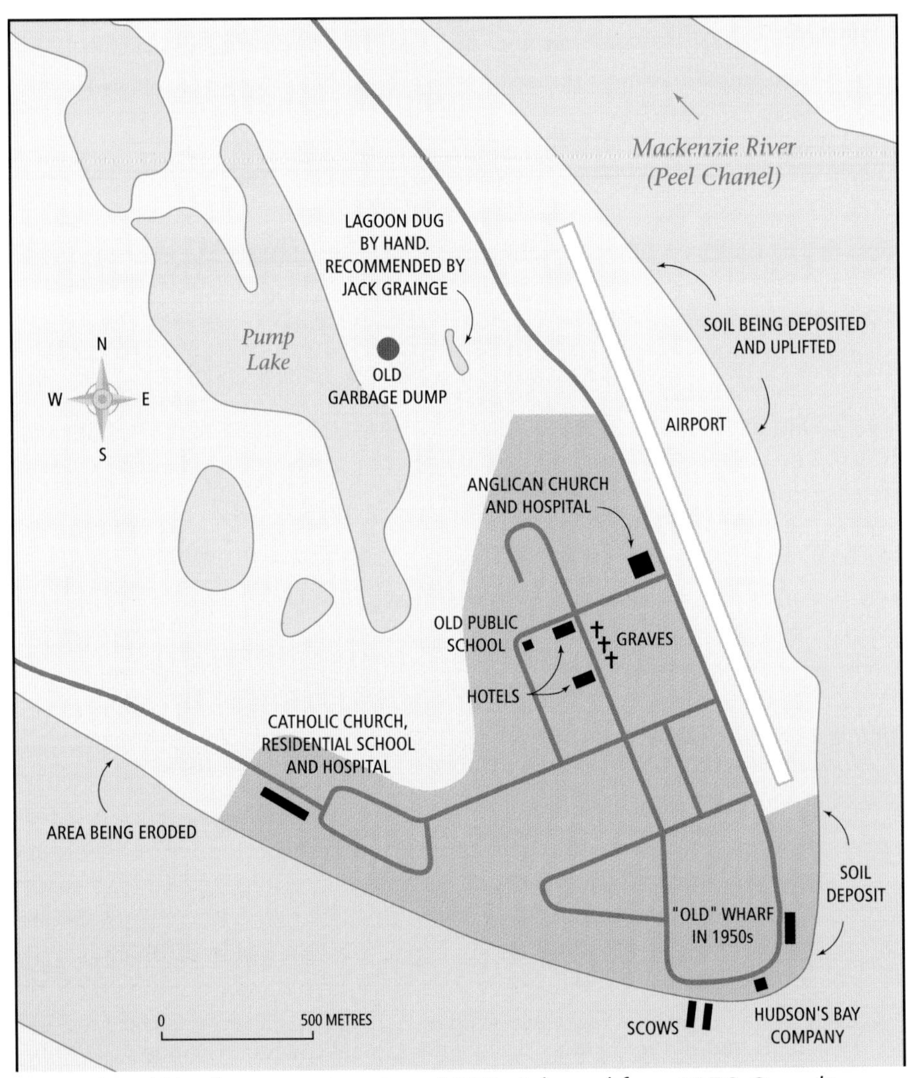

Figure 66. *Aklavik community, circa 1979. Adapted from EPEC Consulting Western (1981), by Johnson Cartographics, Inc. Edmonton.*

However Rutherford was under mental strain. We had faced head winds all the way. Also not being familiar with the route, he flew a long dog-leg, meeting the Mackenzie River about sixty kilometers south of Norman Wells. Consequently the gas in the tank of the Fairchild Husky became low, so low that for the last thirty kilometers Rowdy skimmed the plane along, about five meters above the surface of the Mackenzie River. He told me that if we ran out of gas, he could set the plane down on the water. In retrospect, I do not think that, with the single paddle on board, we could have paddled that plane to the dock. I believe we would have floated down the river. To the relief of all, Rutherford landed and headed straight to the dock.

A few years later, Rutherford was not so lucky. He and R.R. Ross died in a plane wreck in the Yukon Territory.

We passed over Point Separation, where the Mackenzie River begins to divide into delta channels. The Mackenzie River delta is spectacular, one of the world's great deltas. Under a clear blue sky, the delta extended as far as I could see in all directions; winding turbidity-gray, water channels; large and small islands covered with ten-meter-high spruce trees, with alders and willows near some shores; high spruce trees near eroding shores falling sideways toward the water; and patches of spruce trees having fallen pell-mell in all directions and looking like a forest of drunken trees. Those trees had grown only so high and then had been blown down. There were blue lakes in as many shapes and sizes as snowflakes, all of them parts of former channels. What a magnificent sight! A kaleidoscopic panorama that would require an IMAX camera to capture.

Figure 67. *The Mackenzie Delta 1956. Photo by Jack Grainge.*

Aklavik appeared in the distance, a cluster of wooden buildings and a few log cabins on a peninsula, the second half of an S-shaped bend of the wide West Channel, also known as Peel Channel. Circling water was gouging away at the outside bank of the curve upstream of the settlement. It was also gouging the outside curve across the channel from the HBC on the tip of the peninsula. At the same time it was depositing soil along the shore of the community, on the downstream side of the peninsula. Inland from the settlement were spruce trees, a lake and a large, rectangular area which had been cleared of trees. Rutherford said the area had been cleared for an airport. He did not know why the project had been abandoned.

Figure 68. *Aerial photo–Aklavik in flood, 1961. Photo by Jack Grainge.*

Figure 69. *Aerial photo–Aklavik not in flood, 1954. Photo by Curt Merrill.*

Rutherford may not have known much about that airport, but he knew how to land the plane smoothly on the channel. Then, with the engine idling he set the pontoon rudders into the water and maneuvered the plane parallel to the shore, and slightly downstream of the dock. As the plane drew close to the dock, he shut off the engine, stepped out the left door onto the left pontoon, jumped across to the right pontoon and onto the dock. Grabbing a wing strut, he drew the plane gently to the dock.

* * * *

Magistrate Phinney heard the appeal immediately. One evening Dave Jones had hired a teen-age girl to baby-sit their children. A young boy friend came to visit the baby-sitter. They helped themselves to some liquor. Commanding Officer Pat Johnson of the naval station in Aklavik was the justice of peace. He had noted that Jones had not kept his liquor under lock and key, as required by law. He decided reluctantly that Jones was guilty of a crime. He knew that the Joneses were nice, and generous to local people in need. Magistrate Phinney dismissed the case. He ruled that that law did not apply within a private home.

Figure 70. *The wedding of Fred Carpenter and Agnes Pfeffer. Note the muddy ground. Gravel for roads had to be hauled to Aklavik by boat; it was too costly and so this ridiculous situation is the result. Photo by Jack Grainge.*

Dave and Mary Jones were a quiet, friendly, responsible couple, well respected by everybody. He had a steady job and she was the Canadian Pacific Airlines agent. Not surprisingly, when Inuvik became a village, he became its first elected reeve. By coincidence, during World War II, while he had been in training to become a navigator in the R.C.A.F., they rented a second-story suite in a house in Calgary in which I had lived when I was four years old.

The court party took off immediately for Norman Wells.

Possibly thoughts of staying at the Aklavik Hotel, which had one honey bucket for all guests and staff, did not appeal to them. The Canadian Pacific Air guest house at Norman Wells had only one toilet, but it flushed. No matter if passengers had to step across the trickle from the septic tank along the river shore. Undoubtedly they thought that the trickle was spring water.

The naval station had neither ships nor submarines, not even a canoe. It did have many aerials for listening to broadcasts from another distant country. At that time the station's purpose was a military secret.

I checked into the Aklavik Hotel. Guests were lodged upstairs. The washroom which contained a bucket toilet, basin and tub, was at the end of the hall. I think there were about six small rooms. The kitchen and the dining room, separated by a counter, occupied the main floor. Our cook was Rosy Stefansson, about sixteen years old, granddaughter of the famous Arctic scientist, Vilhjalmur Stefansson. He was an Icelander immigrant to North Dakota by way of Gimli, Manitoba. He and his party spent five years with Inuit while exploring western Arctic Islands. He discovered islands off the north coast of Victoria Island, one of which bears his name.

Rosy had a pretty, round face, graced with a perpetual smile. While serving me a cup of coffee, she told me she lived with her parents in town. If they decided to go away and live in the bush, she would go with them. It cost much less to live in the bush because there they fished, trapped or shot most of their food. She was both an interesting conversationalist and a good cook. I was glad she did not leave while I was staying there.

Her assistant in the kitchen was a happy, young Inuvialuit. He made sandwiches, washed dishes, swept floors and emptied the honey bucket into a stinking barrel beside the road. We hoped he washed his hands between jobs.

The first person I met on the street was the personable Don Violette, a missionary supported by the Central Pentecostal Tabernacle in Edmonton. He came north to minister to the Inuvialuit reindeer herders who moved about over a wide territory in the vicinity of Reindeer Station, thirty kilometers north of what is now Inuvik. When Violette's children reached school age, he moved to Aklavik. There he held services in a large room on the main floor of the North Star Inn. I tried to get into one of those meetings, but the place was crowded and overflowing. Local people credited Violette with successfully discouraging excessive consumption of alcohol. The following year he gave me a short ride in his two-seater plane.

Next I visited Dr. Christensen, a friendly old fellow who slopped around the Town in loose fitting trousers and moccasin-style slippers. He had retired after serving twenty-five years in Juliannehaab, near the southern tip of West Greenland. He told me that there were about four hundred people in Aklavik. I told him that I was concerned mainly with everyone receiving clear, uncontaminated water and safe sewage and garbage disposal. With a wink, he wished me luck.

He referred me to Dr. Otto Schaefer, a young, enthusiastic, recent immigrant from Germany. Schaefer was slim and shorter than average. His accent and order of words in sentences was amusing, which taken together with his rapid speech, conveyed warm friendliness and intense interest in the health of everyone in the community.

Schaefer was always in a hurry. With head bent slightly forward, he walked two steps and then ran a step or two. Stretching my long legs I had to move quickly to keep up with him. Dr. Gerry Hankins, who is now writing Shaefer's biography, told me that when Schaefer first arrived in Aklavik the Inuvialuit named him Shik-shik, their name for a chipmunk, an animal with a lurching gait. The Inuit in Pangnirtung, where he served for two years, named him "Lootakooloo," meaning "Little Doctor."

Among themselves, Inuit have names for all outsiders who arrive in their villages. The names are descriptive, but never derogatory.

Once during breakup of ice on the channel, when planes could not land, Schaefer had to perform a difficult emergency operation on a man's knee. Schaefer was not a surgeon specialist, but later a surgeon examined the knee and complimented him on his fine work. It is not surprising that his thorough research on many health problems of aboriginal people has won him high international recognition.

Schaefer's attractive, friendly wife, Didi, invited me to lunch. She served me beef, but Schaefer ate muskrat, for which he had developed a taste. I tried a little of it. It was nothing like anything I had ever tasted, but I liked it. Didi spoke better English than Otto, but Otto often explained to her the meanings of English words that she already knew.

Schaefer took me to Pokiak, across the channel, to see Joe Veitch, a veteran of the Boer War. Veitch was a friendly, old fellow with a shock of dark gray hair, and wearing a dirty parka. He was living on an old-age pension and what he could pilfer while shopping in the HBC store. Whenever the manager chided him, he would put the article back with a remark, "It was

Figure 71. *Joe Vietch. Photo by Jack Grainge.*

worth a try." Previously he had been a laborer in Yellowknife and a laborer and trapper in Aklavik.

Veitch's four sleigh dogs were tethered around his tiny, low-ceiling, one-room shack. His dogs were his friends, and he fed them fish every day. We had to bow down to enter Veitch's door and I had to remain stooped throughout our stay. Veitch had a bed, a table, a few boxes of groceries and clothes, and cats, at least six of them, crawling everywhere.

Schaefer said, "Joe Veitch, the cat is licking your dinner plate." "What's wrong with that?" Joe asked, perhaps with tongue in cheek, "If the cats didn't lick my dishes, they never would get clean."

A few years later, when Veitch was hospitalized, he shot his valuable dogs. He could not endure thoughts of them becoming owned by trappers who might whip them, and when they are not used in summer, feed them barely enough to keep them alive.

Another interesting person was Scotty McNeish, an archeologist working a dig at Firth River near the Alaskan border. He was in town replenishing his supplies.

Bob Smith of the U.S. Fish and Wildlife Service was in Aklavik while waiting out a storm. He flew a Grumman Goose plane, counting geese — an odd coincidence.

Ian MacEwan, a biologist from Ottawa, was studying birds in the delta. He had a store of interesting information about birds in the arctic and the Dall sheep in the Mackenzie Mountains. Later in Ottawa, he married Mary, daughter of R.R. Ross.

Archdeacon Harold Webster and his wife, Edith, lived across the road from the Anglican Church. In 1927 he had sailed in the HBC vessel Distributor to Shingle Point, on the Arctic coast of Yukon Territory, where he taught the Inuit English, reading, writing and religion. In 1928 he established a mission at Coppermine. There he lived, fished, hunted and, with the help of an Inuk and his dog team, visited outlying camps. In England in 1930 he took dental and medical courses for laymen. He then provided medical and dental care along with spiritual shepherding.

The Inuit liked singing hymns. From ancient times, much of their evening entertainment was singing to a steady drum beat, about their hunting experiences. The drums consisted of hides stretched over wood hoops.

Rev. Ralph Gibson, in charge of the Anglican residential school, and his wife, Edna, lived in a house beside All Saints Anglican Church. When I walked by, Gibson was hoeing his huge garden. He took time to listen to me and add his thoughts about the mundane aspects of Aklavik. Rev. Gibson had replaced Canon Colin Montgomery, brother of the Field Marshall of the Desert Rats and later commander of the British Comonwealth Army in Europe.

The church was a small, attractive, wood-frame building. A large painting, 2.5 meters by 1.5 meters, picturing the Epithany hung above the alter. Violet Teague, an Australian artist, pictured that story in a way she thought the Invialuit and Dene, could understand. The Madonna and Child were dressed in ermine, both wearing mukluks. A tall Dene presented a live beaver. An HBC trader offered fox pelts. A kneeling Inuk offered two walrus tusks. An Inuvialuit lady with a baby in her hood held out her hands. A parka clad RCMP with two dogs stood guard. In the background were two caribou and snow houses. Overhead shone a bright star.

The following year when Archbishop Marsh was visiting, Rev. Gibson invited me and many others at the service to dinner at the Anglican Hospital. Archbishop Marsh had lived for years at Arviat (Eskimo Point).

Figure 72. *Anglican Church. Photo by Jack Grainge.*

He, with an Inuit guide-helper and dog team, had run circuits of a thousand kilometers or more to outlying families of Inuit hunters. Sometimes they had to wait out week-long blizzards. I marveled that I was meeting men like him and Archdeacon Webster who had also endured such hardships. I felt as if I were walking on air, however I was just washing the dishes.

I happened across the weed-covered graves of Albert Johnson, marked by a wood cross and Bill England, marked by a tree with his name carved on its two trunks, adjacent to the Anglican cemetery. At that time they were the only graves not allowed within the cemetery, but its boundary has since been extended to include these graves.

Figure 73. *The Mad Trapper's grave. Photo by Jack Grainge.*

News reporters named Johnson the "Mad Trapper of Rat River." He had murdered a man. During the winter of 1931-32, the RCMP search team, based in Aklavik, hunted him by airplane. He was clever and elusive, so they had trouble finding him. He cleverly back-tracked and wound in and out among the trees.

The Sweden-born Carl Garlund, operator of Stan Pfeffer's diesel-electric power plant, became a member of the posse. He did not want a Swede to get away with murder and thereby besmirch the good reputation of Swedish Canadians. The RCMP found Johnson, who lost in the ensuing shoot-out.

England had committed suicide. He and Shorty Wilson had been mining coal at a mine at Shallow Bay, alongside the mainland, eighty kilometers northwest of Aklavik. There seemed to be no reason for him to commit suicide, other than that he might have become "bushed." They had been hauling coal out of the mine which ran horizontally into the side of a small mountain. After they had loaded their barge, he walked a short distance away and shot himself.

Figure 74. *Charlie Gordon dancing in the North Star Inn, 1954. Photo by Merrill.*

Clarence Bell, first president of the Canadian Legion, provided England's body with a suitable funeral. Bell, an RCMP from 1940 to 1943, in the Royal Canadian Navy from 1943 to 1946 returned to Aklavik after he was discharged from the navy. He rented and operated the Aklavik Hotel for a year and then built the North Star Inn, which he opened in 1947.

One evening I attended an Inuvialuit dance in a big room on the main floor of the North Star Inn. A couple of men made makeshift drums. They stretched cotton bed sheets tightly over the tops of round galvanized iron tubs and held them in place by twine wrapped around the rims. The school's Inuvialuit, janitor-handyman, Charley Gordon, was a star dancer.

* * * *

I was surprised to meet R.R. Ross at the three-room school, which had been built about three years earlier. I had previously seen him during the construction of the Fort Smith water system, and at other times socially.

The school had been built without piling, much like many other buildings in the settlement. However, ice-rich permafrost under the school had melted to an uneven depth. Consequently the foundation had sunk unevenly.

R.R. Ross's first step in setting a foundation for the school had been to rip off the floor boards. He then drilled holes in the ground below the floor. He set piles in the holes. In the following years that building foundation was firm, proving that Ross's work had been satisfactory.

Ken Berry, a humorous, solidly built technologist, was steaming holes in the permafrost and setting piles for foundations of federal buildings that were to be built the following year. He told me that many of the buildings in Aklavik were set on piles but many others, including all of the old buildings, were set only on mud sills. Many of those on mud sills seemed to be stable. Some aspects of permafrost engineering might always remain unfathomable.

At the Catholic residential school, I visited Father Biname, a lean and sinewy man of medium height, who traveled by dog team around his parish. He had previously served on the mission ship, *Our Lady of Lourdes,* bringing Christianity, education and medical aid to Inuit in the arctic communities. He showed me the basement of the school. The concrete basement walls were set on concrete footings, approximately 3.5 meters below the ground level. There were neither sags nor bumps in the concrete basement floor.

Dr. Schaefer told me that in summer Brothers had excavated the basement using hand shovels. They had dug the ground to the permafrost. After a few days of thawing of permafrost in the bottom and sides of the excavation, they pumped out the water that had melted and shoveled away the mud. I presume that they dug until the layer of thawed soil remaining at the bottom of the excavation was thick enough to provide adequate insulation to prevent further thawing after the basement was heated. The solid, concrete walls and floor proved that they had done the right thing.

Clarence Bell showed me the lot behind his North Star Inn. It was low-lying land covered with puddles of water. The previous year he had hauled and spread dirt on the grass to a depth of about ten to twenty centimeters. He had done so in order to raise the level of the ground. The soil covered the grass and lichens which had, during the summers of previous years, been dissipating heat by both transpiration and evaporation. In contrast, the black earth absorbed heat from the sun's rays. Thus, the ice in the top layer of permafrost melted and the ground sank. The ground surface became lower than it had been before Bell spread the earth on it.

Stan Pfeffer owned the Aklavik Hotel and the diesel-electric power plant. He told me that the building rested on piles. He was noted for generously wiring electric lines to the cabins of people who could not afford to pay. In winter he erected power poles in the ice on the channel for a line to the few residences in Pokiak. Of course he lost money on that section.

George Roberts, stepfather of Dave Jones, was the Imperial Oil Ltd. (IOL) agent and owner of a tractor and a bombardier snowmobile. He showed me his fish-filled, two-meter-deep refrigerator in the permafrost. He told me that a few days earlier he was fishing as he had been doing for years. That morning an RCMP officer fined him for net-fishing, without having a license. Roberts said that everybody in the settlement fished that way and no one had a licence.

I have always been nonplused by what he told me. Inspector Bill Fraser and his wife were sensible and most pleasant. A year or two later Fraser was hospitalized in Edmonton and his wife stayed with us for awhile when visiting him. Later, he recovered but she herself became sick and died. He then transferred to eastern Canada.

Lee Post, the regional administrator, was a slim, clean-cut man of medium height. His cordial, soft-spoken wife, Lorna, looked like the Hollywood actress, Jane Wyman. The Posts were most helpful to me. Previously he had been superintendent of Reindeer Station, on the east mainland, thirty kilometers north of the place where Inuvik was eventually located. Once while in that position he hired Cliff Anderson to fly him to the Husky Lakes, north-east of Reindeer Station, where he was to make a survey. Due to engine trouble they were forced down. They landed safely on a lake and chugged to shore. They trudged for three days and two sun-lit nights across mosquito-infested, soaking barren land to Tuk. To protect themselves from mosquitos, they wore nets hanging from the brims of their hats, gloves over their hands and wrists, and wrapped the lower parts of their legs with newspapers, the lower ends of which they tucked into the tops of their boots.

When they reached Tuk they knocked on the door of Herb Figures, the HBC Post Manager, and a friend of both of them; they looked like a couple of scruffy, ragged, bearded bums. The friend thought that they were drifters who had worked their way down river. He told them to beat it, but then laughed when he finally recognized them.

Lorna Post told me that their eight-year old son, Wayne, played with two other boys about his age. One of their common games was Cowboys and Indians. The Dene boy could not understand why he could never have a turn being a cowboy.

In winter the people built an enclosed, one-sheet, curling rink. Everyone played, and some of the Dene and Inuvialuit became good, even becoming skips.

Post said that Bill Walls, of the Department of Transport (DOT), Edmonton, had tried to construct an airport to the north of the settlement. He was bulldozing away the trees and surface growths when the permafrost began to melt. His bulldozer sank deep in the muck. Fortunately he had another bulldozer on solid ground outside of the cleared area. He winched the stuck bulldozer out and shipped both to other settlements. The rectangular clearing north of the community was visible from the air for many years. Everyone accepted the conclusion that the construction of an airport at Akalvik was impossible.

Figure 75. *Aklavik during an average flood at break-up, 1972. Photo by Jack Grainge.*

* * * *

Spring arrives earlier in the Liard River's British Columbia and-Yukon headwaters than it does at its confluence with the Mackenzie River. The snow at the headwaters melts quickly. The deluge in the Liard River lifts the main body of ice downstream, breaking it away from the anchor-ice along both shores. The rising water submerges the anchor ice. Thus develops a wide ribbon of unbroken ice the length of the river, bordered by narrow streams of water along both shores.

Soon the lifted ice in the river breaks up. Chunks of ice pile up, damming the river. The river behind these dams rises and rams the huge pile of ice chunks downstream. In

Figure 76. *Photo showing corrugated sheets of iron protecting bank from further erosion, 1972. Photo by Jack Grainge.*

early May the flood reaches Fort Simpson. This half-kilometer-long mountain of ice chunks smashes into the hard ice of the Mackenzie River. If it creates an ice jam, it floods the town. The water level rises eight to twelve meters above normal. The mountain of ice chunks continues snaking down the Mackenzie River. It bulldozes towers of ice chunks upon the outside banks at curves of the river. When the mass of ice and water reaches the delta it spreads out in the three main channels and several smaller channels.

Because of the spread of the water into the several channels, the flood height reduces, being usually five to six meters at Aklavik. The ice jam heaves up a huge pile of ice-chunks on the channel bank opposite the HBC. Occasionally, due to ice jams in some of the channels, more water than usual is diverted into West Channel, causing the water to occasionally rise high enough to flood all of Aklavik, as it did in 1961.

During all spring breakups of the ice in the channels, the fast-flowing water carves away the outside banks of curves. The erosion of the banks at these places continues in summer. The bank at the upstream end of Aklavik faces southwest, so the summer sun thaws the ice layers in the channel bank and accelerates the erosion. During the break-up and the flood that continues for a week or so, slice after slice of the five-meter-high bank slumps off into the channel.

Figure 77. *Father Ruyant and his garden in front of priests' residence. Photo by Jack Grainge.*

In 1953, I estimated the average bank erosion rate at this point to be three meters per year. In the twenty years after my 1953 stay in Aklavik, much of Father Ruyant's huge vegetable garden in front of the mission disappeared. He was the administrator of both the residential school and the Grey Nuns hospital, but he also wielded hoes, shovels and rakes. In 1971 the village administrator placed large sheets of corrugated iron along the channel bank at the bend upstream of the community, which substantially reduced the rate of erosion. Unless someone figures a method of stopping the erosion, the channel will eventually cut through, making Aklavik a small island.

Below the point of the Aklavik peninsula, near the HBC, the water flows more slowly, and consequently the mud settles out of the soil-saturated water. In early years the Anglican mission pumped water for their school and hospital from the channel near their buildings. Each year depositing soil caused mud flats along the channel shore to grow. Each year the operator for the mission had to extend the pipeline further in order to reach the water. Boats could not reach the dock.

After the new shore emerged from the channel, permafrost developed. Ice layers in the ground formed and grew. The ground rose high and dry creating an airport for small planes. Mother Nature had accomplished what Bill Walls with his bulldozers had failed to do.

* * * *

Lee Post operated a summer, community water supply system. The source of water was Pump Lake. During the annual spring flood, water from the Channel, polluted by sewage from both missions, flowed into the north end of the kilometer-long, Pump Lake. From there water was pumped directly to a summer water distribution system. This system consisted of above-ground galvanized iron pipes, feeding to hydrants and all buildings with household plumbing.

Most of the bacteria polluting Pump Lake would have become attached to silt particles in the water, which after two or three weeks, settled out. However, the water being cold, a few of these bacteria might remain viable for a few weeks.

After the channel flood had passed and the turbidity settled out, water from Pump Lake was pumped and piped throughout the community. In winter the pipelines were drained by disconnecting a few of the joins. Lee Post and I agreed that they should cover the open ends of the pipes to keep the pipes clean.

The water was not chlorinated. After discussions with Stan Copp, my boss, I did not suggest chlorination. The water flow-rate varied according to the demand. Therefore a costly, variable-feed chlorinator would have been required, and it would have been too difficult for the workman in that community to operate. The installation of a constant-feed chlorinator would have required a large, treated water storage reservoir, which would have been difficult to maintain in a clean condition. Believe me, I have seen many, many filthy water storage tanks in northern communities.

The following year Dennis Prevost, assistant to Lee Post and operator of the water supply system, installed a diatomaceous-earth filter. This filter consisted of a bed of pulverized, diatom particles. It removed the insects and plankton. It could not be used while the water was turbid, because the silt would have quickly plugged the filter. This type of filter had just come on the market. At that time it was the best filter available for removing plankton from clear water.

Prevost gave me a ride, seated behind him on what he called a motorized toboggan. He and Moose Kerr, the public school principal, had replaced the front wheel of a small motorcycle with a short, narrow toboggan. We ploughed through thirty centimeter deep snow without difficulty. It was a forerunner of a skidoo, and Mr. Bombardier did not give Prevost credit.

Later Lee Post hired Nels Hvatum, a local Norwegian trapper and hunter, as the water system operator. In later years, Hvatum gave the job to his young son, Herschel, who did the job well. I visited with the Hvatums a few times. A few years later, Maggie, Hvatum's Inuvialuit wife, suffered extensive third degree burns while escaping from a fire in their log cabin. Herschel suffered some burns and two children died. Daughter Nellie was not hurt. Later Herschel attended university in Edmonton and became a highly respected engineer.

When she grew up, Nellie Hvatum married and became famous as Nellie Cournoyea. Years later as manager of the CBC station in Inuvik, she kept the station on the air throughout a Canada-wide strike. She organized the Committee for Original People's Entitlement (COPE), becoming its spokesperson and effectively its manager. Later the name was changed to Inuvialuit Regional Corp (IRC). Cournoyea was elected and served sixteen years in the NWT Council, serving as chair from 1991 to 1995.

The Catholic Mission had plumbing for running water and sewage in their school and hospital. They had both a water reservoir and a sewage tank in the basement of their school. They pumped water from the channel into the reservoir, except when they connected to the public, summer water distribution system. At intervals they emptied the sewage tank to the channel.

During spring and fall the Anglican Mission piped water through a long pipeline on stilts to a reservoir. During summer they connected to the public, above-ground water distribution system. In winter they put blocks of ice in the reservoir. The two missions informed each other so that the Anglican Mission would not be drawing water while the Catholic Mission was discharging sewage.

In winter, people cut ice from the channel, upstream of the settlement. They stored the ice blocks in piles outside their houses. Sometimes they stored them in the forms of walls protecting their doorways from winds. The water was contaminated somewhat by the men handling the ice blocks. The water could have been chlorinated by adding Javex when the ice was added to the water barrel. I did not suggest doing so because adding, mixing and testing would have been complicated. The men cutting and handling the ice blocks were local men who would not usually be exposed to microorganisms foreign to the community. Usually people drinking the water would become somewhat immune to microorganisms of local workmen.

Straying from the application of standard rules of sanitation may sound odd now. Stan Copp and I made such decisions after consulting with the doctors and nurses. If we had not relaxed sanitation requirements, we would have had to close almost all of the restaurants and water supply and sewage systems in the NWT. We recommended changes which could be afforded. Even so DI&NA engineers often criticized us for being too strict. "Impractical" was their expression. Often we would make recommendations for minimal improvements, which they would not accept. We had to strike an acceptable middle course, different for each community. For many years Department of Indian and Northern Affairs (DI&NA) was the only administration with enforcement power in municipal matters. In the 1960s, the Northern Health Services, DNH&W, became organized with Dr. Gordon Butler as the regional director. He seemed to share authority in this respect with Stewart Hodgeson, Director, NWT Administration. They became close friends.

* * * *

Two motor boats with low cabins were tied up to the shore opposite the HBC store. Inuvialuit families living on them carried water from the public hydrants. They discarded their wash water, but not their sewage, overboard. I thought there was no reasonable alternative.

Aklavik residents threw their wash-water in their yards, resulting during rains in the road ditches becoming sewers. Stan Copp and I thought that any suggestion to do otherwise would not be accepted. Considering budget constraints in those days, we did not recommend a piped sewer system. We thought that because of the probably uneven heaving and settling of the ground due to the permafrost, a subsurface system would be impossible. Also we thought above-ground utilidors would be too costly and too disruptive to the roads, yards and playgrounds.

Pit outhouses had been tried, but the pits filled with water which turned to ice in winter. Then rains and melting snow spread the pollution.

Plastic bags for lining honey buckets had not yet come on the market. People used honey buckets, and emptied them into barrels on stoneboats at various locations near the roads. At regular intervals a man with a small tractor dragged the stoneboats to the dump. The hotel had its own barrel on a stoneboat by the road.

The summer dump was almost a kilometer north of the community. The workmen tipped the barrels containing toilet sewage over as close to the dump as they could get without getting their boots or barrels into the mess from a previous dump. Consequently, as the summer progressed, the dump grew closer to the community. The flies were so thick that coming to the dump felt like bumping one's face into a thick curtain.

I suggested to Lee Post that he experiment with a small sewage lagoon. From my department's fund for research projects, Lee hired laborers to dig by hand a half-meter-deep sewage lagoon. It filled with water from melting ice layers in the soil. The toilet sewage was then protected from flies, and the sewage odor was eliminated. People spoiled its appearance by throwing tins and other non decomposable trash into it. It was a good system as long as the community remained small, but it required better care.

In winter the people dumped their trash and barrels of frozen sewage on the ice in the middle of the channel opposite the lower end of the community. Stan Copp and I surmised that during the spring floods, the ice would jam the mess all the way to the sea. During the ice jamming, no one downstream would use the turbid water.

Most of the trash would settle in the accumulating layers of channel-bed silt. Some junk might float to various beaches, but the ice chunks would have broken them up and the roiling water would have washed them clean. After plastic bags came into common use, this practice would not be acceptable. Plastics could choke fish and the plastic bags of garbage and sewage, that could drift onto downstream beaches, would be unsightly.

In our opinion, the vast Arctic ecosystem was capable of digesting the wastes. The water in the western Arctic seas, having originated in the Arctic Ocean and from north-flowing rivers, is nutrient poor. The sea life would benefit, ever so slightly from the organics in this garbage and sewage. In contrast, the eastern Arctic seas are rich in organics. They benefit by upwellings of nutrient-rich water originating in the tropics. Those upwelling waters support huge numbers of seals, whales, narwhales, fish and birds.

* * * *

Before leaving the settlement, I discussed with several people the pros and cons of moving Aklavik to a place on the mainland. Many of them agreed and suggested that they thought a suitable site was alongside the Husky Channel, just twenty kilometers to the west. Most thought it was a good idea but it was coming too late. Too much had been spent in large buildings already in Aklavik, and a big Catholic school then being built. When, with strong arguments, they ridiculed my idea, I laughed along with my critics.

In my report I recommended moving the settlement to the mainland. I supported my opinion with arguments about the community eventually becoming a large town, there being no possibility for an airport accessible by road, no possibility for subsurface piped water and sewer systems, no nearby source of gravel for roads, the high cost of building foundations and the fact that the Peel Channel was eroding the nearby bank west of the community.

Of course the officials in Ottawa were previously aware of the difficult conditions in Aklavik. They needed a townsite to be an administrative center, with a substantial airport and a boat dock on a navigable channel. Evidently my report confirmed their opinion and provided them with supporting information. The Cabinet accepted their recommendation.

Soon after the announcement on 23 December 1953, that Aklavik would be moved, a Time magazine reporter flew on the weekly Edmonton-Norman Wells, DC3, mail run, intending to reach Aklavik. When he realized that if he went farther he would have to spend a week in Aklavik and Norman Wells, he returned to Edmonton on the next day's flight. Apparently he asked people flying south from Aklavik about the place. I suppose a wag among the informers told him that Stewart (Stu) Hill was known as the Angel of the North, and that is what the reporter reported.

Hill flew the CPA single-engine, Norseman airplane from Norman Wells to Aklavik as well as to other settlements in the region. He was a hardworking, reliable pilot, but sometimes un-angelically impatient with tardy passengers. Probably he was also unhappy about being separated from his family except for a month during spring break-up and another month during winter freeze-up.

People did not really know him. When he was in Aklavik, he boarded with Dave and Mary Jones, who liked him. However when Mrs. Norris opened a boarding house, Mary suggested that Hill would be more comfortable if he moved there. He refused. She said that their baby's crying probably disturbed him. He replied, "No. I love the baby and I like to baby-sit for you. I will not move." That was that.

* * * *

In the years 1958 to 1960, the Navy base, the hospitals and residential schools of both missions moved to Inuvik, but did not destroy their buildings in Aklavik. The missions were compensated for losing their schools and hospitals. The government operated the day school, but both missions built and operated residential schools for out-of-town students. The merchants and hotel owners were also compensated so that they could build better facilities in the new town. After paying compensation, Northern Canada Power Commission (NCPC), bought the deisel-electric power plant in Aklavik.

Many people moved to Inuvik with the promise of better houses. Northern Affairs Minister Jean Lesage understood the various interests involved, and how to spend funds wisely. He was able to avoid most local objections to the move. His administration planned that Aklavik remain only a small fishing, trapping and hunting community.

More people than expected remained in Aklavik. They continued trapping, hunting, fishing, making and selling fur garments and handicraft articles, working weeks-on-and-off at gas and oil well drilling sites, and some receiving welfare and pensions. Moose Kerr, the principal of the school, crusaded to save the community even though its staying was a foregone conclusion. He, together with others, composed a song, Aklavik Will Never Die, which his students sang at concerts, and some people sing even now.

In time a new school, health clinic, fur garment and handicraft workshop and houses were built in Aklavik. Unfortunately in 1970 the Anglican church burned down and the beautiful painting by Violet Teague was lost forever.

In 1961 DI&NA hired Norm Lawrence, President, Associated Engineering Services Ltd., to make recommendations for improvements to the water supply and sewage and garbage disposal systems. He recommended the construction of wood-frame containers on stoneboats. Each container should contain two barrels, one for trash and one for toilet wastes. As with the existing garbage-sewage barrels on stoneboats, they should be located at convenient places around the settlement. The contractor would continue to haul them to the garbage disposal point. One of his main recommendations was that they should not construct a year-round water distribution system.

References:
Jones, D. 1996. Personal communication.
Jones, M. 1996. Personal communication.
Lawrence, N. 1996. Personal communication.
Pich, M. 1996. Personal communication.
Schaefer, O. 1996. Personal communication.
Webster, Rev. J.H. ca.1988. *Arctic Adventures.* Ridgetown, Ont.: G.C. and H.E. Enterprises.
Zubko, D. 1996. Personal communication.

INUVIK

On 23 December 1953, the Edmonton Journal headline read "AKLAVIK TO BE MOVED." At Aklavik, Father Biname stopped work on a new residential school. The search for a new site began.

An ad hoc committee in Ottawa decided that a replacement community for Aklavik must be (1) on the mainland, (2) on a navigable channel on either side of the Mackenzie River Delta, (3) accessible by road to the not yet planned, NWT roadway system, (4) large enough for a town of an eventual five thousand population, (5) have a site for a substantial airport, (6) have a source of good water, (7) allow for an acceptable sewage treatment and disposal system, (8) have a source of gravel for roads and airport runways, and (9) have subsoil that should preferably not be silt, which is subject to frost heaving and serious settling upon thawing.

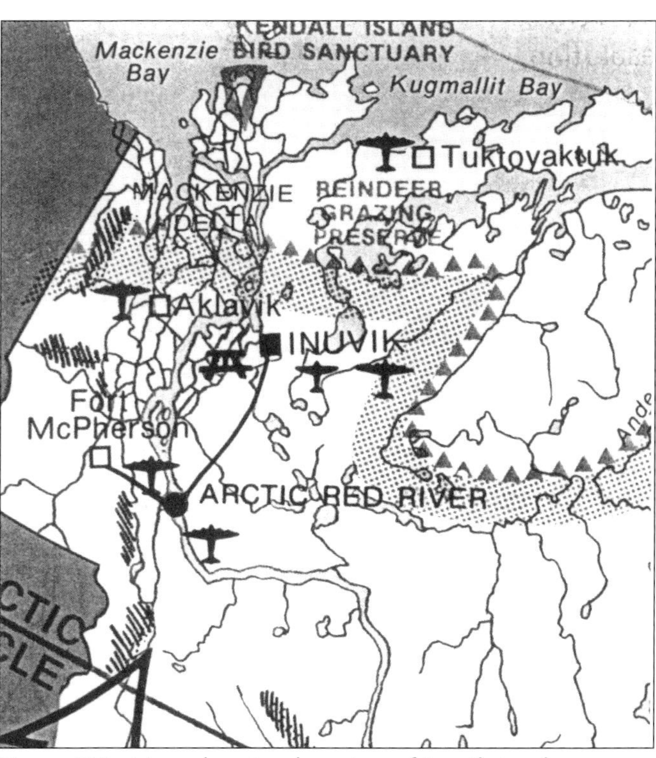

Figure 78. *Map showing location of Inuvik 'replacement community'. NWT Travel Map (1976).*

Curt Merrill, an enthusiastic, 37-year-old geologist, who had served in various positions in the Department of Indian and Northern Affairs (DI&NA), Ottawa, headed the search party: from Ottawa there was John Pihlainen, Hank Johnson and Roger Brown, young engineers, Permafrost Section of the National Research Council (NRC) and Keith Fraser, geographer, Dept. of Lands and Technical Surveys. From Edmonton, Ken Berry, technical officer, Department of Public Works (DPW), and I rounded out the team.

Dan McLeod, retired laborer, trapper and fisherman from Aklavik, became our competent cook. Leffingwell, an Aklavik Inuk, grandson of Dr. Leffingwell, a geologist in Vilhjalmur Stefansson's high-Arctic exploration party, was our adviser.

He helped set up camp and trounced all of us playing crib. Fred Norris transported our equipment and the crew by tractor-drawn sleigh to the Husky Channel site, twenty kilometers west of Aklavik.

Don Landells, Fox Helicopter Services, Edmonton, flew a two-person Bell Helicopter that we used for our transportation to the various sites. The large number of personnel involved was a reflection of the high media attention given to the planned moving of a far-north community with a quaint name.

Everyone but me arrived in March and April 1954, and set up camp at the Husky Channel site. They scouted by helicopter the several possible mainland sites on the east and west sides of the delta. They rejected all but the Husky Channel and East Three sites.

RCMP Inspector Bill Fraser recommended an additional site, north along West Channel from Aklavik. We nicknamed it Fraserville. Later in the summer, Johnny Pihlainen, Hank Johnson and Roger Brown drilled test holes there that showed ice layers. The surface soil was boulders and clay, with no gravel for road and airport construction. They rejected that site.

Figure 79. *Jack Grainge disembarking from Bell helicopter at base camp, Husky Channel. Photo by Curt Merrill.*

I arrived at the camp in mid-May to find everyone complaining about diarrhea. Dan, our cook, felt that he was getting oblique looks. I sent away samples of water from Husky Channel and all lakes that might be suitable sources of water for the new town. In a few days, I received a telegram from Stan Copp, my boss, relayed from Aklavik, "SULFATES (Epsom salts) IN DRINKING WATER FIFTEEN TIMES ALLOWABLE LIMIT." From then on we carried water from the channel and the cook was relieved. So were the rest of us.

Landells was a skillful helicopter pilot, as he proved many times. Once with me as his passenger, he flew low above a Navy, four-person snowmobile in which some joy-riders were speeding along on the channel ice. One man, standing with his head and shoulders sticking through the open top of the snowmobile, was waving to us. Don flew above them at their speed, and gradually dropped until the helicopter was less than a meter above them. The lookout snapped his head in fast.

Another time, with Berry as passenger, Landells skimmed above the Peel Channel ice, chasing a wolf. Flood water had previously lifted the main body of ice on the channel, breaking it away from the submerged anchor ice along both shores. Because there were three to four-meter-wide strips of open water along both banks, the wolf could not swerve fast enough to escape from the

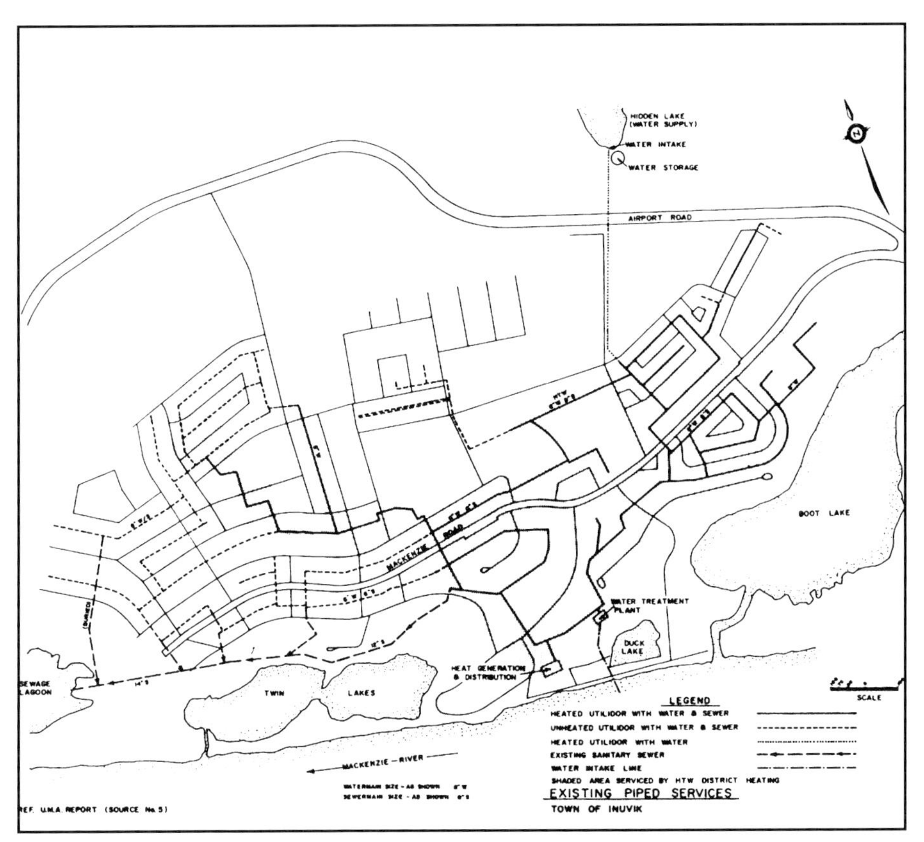

Figure 80. *Map of the town Inuvik, from EPEC Consulting Western Ltd. (1981).*

wide ribbon of ice along the channel. Landells flew behind the wolf running a landing strut between his hind legs, flipping him over and flying past him. The wolf jumped into the water, swam to shore and disappeared into the forest.

Landells must have picked up his skill from his famous boss, Tommy Fox. One time while landing on a lake east of Fort McMurray, one of his pilots had sheared a three-meter length off the end of a wing. Fox flew there in another plane. To restore the wrecked plane's balance, he sawed an equal amount off the end of the other wing and flew the plane to Edmonton. The airplane safety inspector in Edmonton decided that that was dangerous flying and took away Fox's flying license. Fox went to Ottawa where he explained the circumstances to the airplane safety director. The director congratulated Fox for his ingenuity and gave him back his license.

Years later Landells started his own helicopter service, specializing in transporting wildlife biologists studying endangered animal species. Biologists counting mountain sheep in both United States and Canada preferred Landells as their pilot because he was both careful and skillful in skirting the cliffs. However in 1986, while flying biologists, who were counting bighorn sheep in California, his helicopter was caught by a wind gust and crashed on a steep mountain ridge. He and one biologist died. Two escaped with minor injuries.

* * * *

Both the East Three and the Husky Channel sites were on the mainland, bordering navigable channels. Both sites were sloped and well above flood level, making possible good drainage and the operation of a system of shallow subsurface sewers. When warmer weather arrived in June, the NRC men drilled holes that showed substantial ice layers in the soil throughout the Husky Channel site.

Landells flew me to Christine Norris's trapping cabin on an island bordering East Channel, twenty-five kilometers south of East Three. A widow of an independent trader, Adolphus, she was a pleasant conversationalist, excellent cook and efficient trapper. Two years later when I stayed overnight at her boarding house, I found that she was also a clever business woman. I treasure my memory of her.

Peter Thrasher and another young man trapped for her. Peter was the son of an Inuvialuit mother and Billy Thrasher, the Polynesian captain of the Catholic Mission's ship, *Our Lady of Lourdes,* which sailed the arctic seas.

Christine Norris was the mother of Fred, who transported our equipment from place to place; of Danny, who later served for a few years as commissioner of the NWT; and of Clara, wife of Barney McNeil. At the time McNeil was the weather observer at Aklavik. He had come to Aklavik in 1948 as a telegrapher and weather observer in the Royal Canadian Corps of Signals. Later he became airport manager at Inuvik and eventually justice of peace and coroner. He had achieved local fame by swimming to safety after he, driving his tractor, had broken through the ice on the channel. Previously he had been a competitive swimmer in Montreal, his home town. Even though being on the water all their lives, local people had no opportunity to learn to swim. In most places the water was usually too cold for swimming.

During sunny summer days in Aklavik and Inuvik, the water does become warm enough for swimming during sunny days in July. One Sunday while the survey crew was there, I cut out a piece of plywood about a meter by a half meter, on which we took turns water skiing.

The previous year, Danny Norris, had been a patient in the Camsell Hospital in Edmonton, in the final stages of recovery from tuberculosis. I visited him and invited him to our house. Later, during a few days of my vacation, I was surveying for Don Stanley, in what soon became the town of Sherwood Park, east of Edmonton. I hired Norris as my helper for a few days. Years later he told me that these had been his happiest days in Edmonton.

In May I measured the flow under the ice in East Channel. With a needle bar I chipped through the ice. It was one meter thick where it was covered with about thirty centimeters of snow, and a half meter thicker where the ice was wind-swept. In late winter there was no flow. Obviously the channel freezes to the bottom at upstream shallows.

* * * *

By the time I had finished my work, I was feeling the early symptoms of conjunctivitis. I wore clip-on, dark lenses, but light was refracted into my eyes by the top edge of my rimless glasses. To save my eyes I wrapped black electrical tape around the upper edges of my glasses. To heck with appearances, I felt better.

On the first flight after the West Channel break-up on June 4, I left for Edmonton. Lee Post, the town director, was a friendly, informative flight companion. During a stopover in Yellowknife, we visited Bert Boxer, who was working for DI&NA. Boxer had previously been a trapper and then an employee of the Anglican Mission.

I returned to Aklavik in late June on the same flight as Ted Garrett, Technical Officer, Department of Transport (DOT), Edmonton. His job was to look for an airport site. With his spontaneous wit, he added a great deal to the humor within the party.

The survey party had already moved their tent camp to East Three. They had rented Fred Norris's small, aluminum barges. The resourceful Merrill had made a raft with spruce poles and plywood riding on empty oil drums. Merrill had pulled these with Jim Macdonald's tug boat.

The site met many of the criteria set by the Ottawa committee as being necessary. It was alongside the mainland shore of East Channel, navigable from the Mackenzie River. It could connect to a future NWT road network. It was a large, slightly sloped shelf, high above flood water level of East Channel.

Along the shore was a half kilometer-wide band of attractive stands of birch amid Labrador tea and other tundra flowers. It contrasted with the nearby arctic barrens to the east. Unfortunately much of the site was underlain by ice-rich soil, some of which melted during the summers.

Richard Ethier, my Métis helpmate, was the son of Con Ethier, an ex RCMP, who had bought his way out of the service in order to get married. In those days an RCMP was not allowed to marry before he had completed five years of service, had a thousand dollars in his bank and had permission to marry. Richard and his sister were strong. She was a professional wrestler in Edmonton.

Once while Richard and I were walking along on the apparently dry tundra, he joked, "If you're thirsty take a drink." The nearest lake being a long distance away, we both laughed. However he bent over, brushed aside some lichens, swept his cupped his hand along a surface of exposed ice scooping up a palm full of water – excellent flavor. Ethier said that travelers across the tundra in summer depended on water like this.

Much of the flatland on some tundra bordering Inuvik was soaking in water five to fifteen centimeters deep, and dotted with forty centimeter-high hummocks. We hopped from one to the other of these knobs without stepping in the water.

I asked Ethier, "Do you realize Richard, this land is a desert? I am reading a book that says so." He joked about taking me to see a doctor. I explained that geographers call it a polar desert because the annual precipitation is low. They do not take into account the fact that because of short, cool summers, the annual evaporation rate is low. Fortunately for us, Merrill had supplied us with flexible rubber mukluks for walking through this soaking desert.

The soaking tundra was ideal breeding grounds for mosquitos, black flies and bulldogs (horse flies). Merrill gave us mosquito nets. We wore them over brimmed hats and they hung down to our shoulders. Our rubber muklucks protected our ankles. Leather gloves protected our hands and wrists.

For transportation to Aklavik we used our own motor boat or hired Mike Zubko to fly us in his Cessna 180. With a grin as wide as the Mackenzie River, he would greet us, "What are you caricatures doing now?" The distance was fifty-five kilometers by plane and seventy kilometers winding through the delta islands by boat.

The first investigators, studying the aerial photos, identified an exposed glacial deposit of gravel, a kame (an isolated glacial deposit of gravel), north of Twin Lakes. Drilling by John Pihlainen and his crew showed that much of the rest of East Three site was silt interspersed with thick layers of ice.

* * * *

The obliging Richard Ethier and I set out for the nearest lake southeast of East Three. We carried a paddle and a rubber raft for me to take soundings. Studying an aerial photo, we walked through the light forest in what we thought was the right direction. After a while we realized that we had missed the lake, but did not know by how much, nor on which side. When eventually we stumbled upon the lake, we toasted our exploratory skills with clear, cold water. We were lucky that there was only one lake, otherwise I would probably have sounded the wrong one. On our return, Garrett claimed that the reason for us going astray was that I had wandered through the trees carrying our paddle sideways in front of me. He claimed that the only way I could move was sideways and that I had sidewaysed my way around the lake. We named it Hidden Lake and surprisingly the name stuck.

During the study of the aerial photos, Merrill named Boot Lake for its shape and Twin Lakes for the obvious reason. While Ethier and I roamed about we came upon the only unnamed lake. For this Ethier suggested the name Duck Lake. Unimaginative though it was, that name also stuck.

Figure 81. *Aerial photo 1960. Sewage lagoon in foreground, Twin Lakes mid-photo, and Boot Lake in distance. Photo by Jack Grainge.*

I found that much of Hidden Lake was more than six meters deep and a small part of it more than nine meters deep. I calculated the approximate volume of the lake, and concluded that, with surface inflow from higher ground, it would be large enough to serve the town in summer for many years.

Figure 82. *Water intake under construction in Hidden Lake. Photo by Jack Grainge.*

The water in Hidden Lake is soft and clear. Micro-straining to remove macro-organisms, followed by chlorination would be the only treatments required. The straining was provided by a newly invented micro-strainer. It consists of a revolving, cylindrical, stainless steel, fine screen, half immersed in the incoming water. Larger plankton get caught on the screen and further constrict the spaces so that the tiniest of plankton are removed. Before re-submerging, an adjustable fine spray of water washes some, but not all, of the macro-organisms off the screen and to waste.

In winter, after surface ice forms on East Channel, the turbidity particles settle, leaving the water clear. The surface ice protects the water from bacterial pollution, but chlorination of surface water supplies is standard and therefore was required. When the town grows too large for Hidden Lake to be the only source, clear water from the channel could be used in winter. That clear winter water could also be pumped to refill Hidden Lake.

* * * *

The party returned to Aklavik to await the arrival of the Honorable Jean Lesage, Minister of DI&NA, R. Gordon Robertson, Deputy Minister, and L.A.C.O. Hunt, Administrator of the western NWT. We visited the local people and played card games. I was luckier than I had ever been. Winning was more fun. Also Pierre Berton, at that time, the Editor of McLean's magazine, spent a couple of hours talking with us.

When the Honorable Jean Lesage and party arrived, they flew to East Three, and looked around. Later they met with us and said that they agreed with us that East Three was a satisfactory site. Although we had not found a large, airport site, aerial photos showed that a site in the area could be found. Lee Post invited them to a dinner and an evening at his home.

The next morning Robertson informed Merrill that the Minister decided that the move to East Three was on. He appointed Merrill manager and told him to start preparing the site. He added, "By the way Curt, some barge loads of buildings and supplies are already on the way. Wire destination instructions to NTCL."

Swanson Lumber Co. had already shipped several million board feet of spruce lumber from their mill at the south end of Wood Buffalo National Park. Also DI&NA employees had knocked down and shipped several small houses in Yellowknife, which had been used by shipping transfer crews during the construction of the Canol pipeline and oil wells at Norman Wells. The planners in Ottawa had also bought and shipped a bulldozer, a backhoe and several trucks. That bold decision to send these supplies cut a year off the construction time of the new town. It also maintained the momentum of the community move. By this time everyone involved realized that Aklavik would continue to exist. East Three would not be a new Aklavik. It would become a new regional center.

Merrill lost no time. He hired Adolph Koziak, a local trapper, as foreman. He hired Fred Norris to bring his D2 Caterpillar tractor and other materials from Aklavik in his barges towed by his tug, Barbara Jean. Fred owned barges that the USA Army had left when they abandoned construction of the Canol Pipeline. Merrill hired him to strip the tundra off a hectare of gravel. By freeze-up it had thawed and the water had drained away, promising a supply of gravel for road construction the following year.

In September Dick Snowling and his assistant, Dick Bower, Dept. of Energy Mines and Resources, surveyed the area. They produced a large map with one-foot contour lines. Two years later, Charlie Gordon, not related to the Aklavik man of the same name, made a legal survey of the lots.

In late September, Merrill closed the camp. He paid the local laborers with chits at the HBC store of Aklavik. Herb Figures, the manager, became banker for all the workmen. Mike Zubko flew Merrill and the surveyors to Norman Wells in his Cessna 180. The regular Norseman service had been shut down during freeze-up of the channel.

* * * *

For the 1955 construction, Merrill chose as his engineer the friendly, solidly built Charlie Walrath, DPW, Edmonton. Walrath was always a lighthearted, interesting conversationalist. He was a minister in the Reorganized Church of Jesus Christ of Latter Day Saints, but he never talked about religion. Soon after the job had started, he prepared a petition requesting the government to make East Three liquor free. He signed it, as did most of his employees and many workmen employed by contractors. The prohibition was generally effective, and proved to be well worthwhile.

After retirement Walrath spent his spare time working for aged church members who could not afford to pay for repairs to their houses. He charged one dollar per hour, compared to the $15 hourly tradesmen's rate. He did good work. I hired him to help me rebuild a chimney, but I paid him bricklayers' wages.

During May 1955 Merrill and Walrath returned and again hired Adolph Koziak as foreman. Koziak remained in charge and later became a town employee. Eventually he retired on pension at Campbell River, B.C.

Merrill expanded his crew to more than a hundred, approximately thirty of them local men. Northern Transportation Co. Ltd. delivered road-building equipment consisting of a D6 Caterpillar bulldozer, a two-yard backhoe that could be converted to a dragline, about seven GMC, five-ton, dump trucks, a steam jenny and a Bombardier snowmobile.

Using a local bank account, Merrill bought materials and paid wages to local laborers. Peter Thrasher operated the bulldozer. In 1954 he and others of Merrill's crew had taken lessons in operating earth-moving equipment at a school at Leduc, Alberta, managed by Bert Boxer.

* * * *

Merrill's camp consisted of kitchen, dining room, sleeping quarters, washrooms and offices in small houses called 512s. Mary, his pleasant wife, became the efficient, hard-working chef.

The floor area of the 512s was of course 512 sq. ft. (46 sq. m). They would later be moved to residential sites and sold to the people who had accepted compensation for moving from Aklavik. These houses were small, but they were larger than the cabins from which the occupants had moved. They were larger than log cabins in which our family and many other homesteaders in northern Alberta lived during the 1920s and 1930s. The main part of a 512 was the kitchen-dining-living room. At the rear of the house were two small bedrooms. At the front of the house was an indoor porch and a small room containing a honey bucket.

Merrill set the camp on the brow of Duck Lake. Using a 300 mm flexible plastic pipeline (a new product then), he siphoned water to this camp from Hidden Lake. During the summer, some wags fired twenty-two bullets into the pipeline, producing mini-fountains. Taking advantage of the pressure-head provided by the elevation of Hidden Lake, Merrill maintained a 2250-liter tank full of water in a building beside his kitchen. He installed a pressure water distribution system for his kitchen and camp washrooms.

He piped the sewage by gravity to Duck Lake. The discharge was innocuous, so much so that two families set up tents on the shore, not far from the sewage disposal point. They moved when the RCMP told them they were using sewage-contaminated water.

Engineers of the Edmonton office of the Department of Public Works designed and supervised the construction of the wharf.

DI&NA hired a consortium of engineering companies headed by Foundation Engineering Co., Montreal, later renamed FENCO, to prepare plans for the town, keeping in mind that the population might eventually rise to five thousand. They also made proposals for water supply and sewer systems. During the summer of 1955, their engineers, surveyors and architects, as well as foundation engineers from R.M. Hardy and Associates, Edmonton, made studies of the site.

FENCO designed the town with cul de sacs. In our written comments, Copp and I pointed out that water distribution and sewer systems to serve those cul de sacs would be too costly. They did not agree, but made minor changes to the plan. The planners returned the following year to site the power plant, schools, students' hostels, hospital, hotels and stores.

* * * *

In order to make suggestions regarding foundations for buildings and roads, John Pihlainen and his crew transferred their study of permafrost from Norman Wells to Inuvik. Their study was a follow-up to one in 1951 by my supervisor, Stan Copp et al. From 1954 to 1962, Pihlainen measured ground temperatures around and under roads, large and small buildings, the storage tank, the power house and air strip. Dick Hill, the manager of the science laboratory at Inuvik, helped them. When Inuvik became a town, Hill was elected its first mayor. The previous year Dave Jones was elected the reeve of Inuvik village.

To lay a solid foundation of gravel for the gravel roads, Walrath's construction crew tried stripping away the organic overburden consisting of lichens and silt. This created a three meter-wide, boggy trench of mud and water. Switching to laying willows on the un-stripped ground crossways of the road and dumping gravel on top was no less a headache. Besides there would not have been enough willows. Finally they simply dumped almost a meter thickness of gravel on the natural ground. If sags developed, they filled them with more gravel. They had found that to be the only practical construction technique. The roads have stood up well, but were dusty.

They constructed roads from the gravel source north of Twin Lakes, to the camp at Duck Lake, and a loop around the site with spurs to the wharf and south of the site. There they discovered another large deposit of gravel, the top of another kame.

During road construction, FENCO engineers were making basic studies of terrain conditions. Many roads had been constructed before the recommendations based on these studies arrived. Also culverts arrived after Walrath had made some by welding oil barrels together. These make-shift culverts served well during construction. With a tug and barges, Fred Norris salvaged a large number of oil barrels from an abandoned LORAN navigational station at Kittygazuit about 80 km north. Also he brought back some of the steel from the 214 m-high, LORAN radar tower that had been knocked down.

In June 1956 Slim Semmler built a store at East Three. It looked like a typical HBC store, white with red roof shingles and a large front window. Undoubtedly, Herb Figures, the manager of the HBC store in Aklavik, heard about it. In August he built a store in Inuvik.

Walrath's crews built a large workshop with space for repairing and parking heavy equipment. Its concrete floor was built on a gravel pad, but that was insufficient to prevent the thawing of the ice layers below it. When I was there two years later, the floor had sunk to a deep, dish shape.

They erected five large warehouses, four of them on piling and one, as an experiment, on a thick gravel pad. All five of them stood up well. The piles were cut from local spruce.

They sank piles for forthcoming large buildings and pipelines, and in the process learned new techniques. They jetted steam down through a 20 mm steel pipe to a depth of 4.3 meters. Then they set a five-meter-long log, butt down, into the hole and rammed it home with a pile driver. They found that the best time to set piles was May or June, when the ground temperature at 3.7 meters depth was at its annual low, so the pile froze quickly at its base. They found that the piles set in September froze first at the top, and consequently sometimes they later heaved upward.

Figure 83. *Sinking piles into the permafrost. Photo by Jack Grainge.*

Merrill let contracts to various people with power boats to cut logs in the delta and drag them to the East Three wharf. The price delivered at the dock was ten dollars for five-meter-long piles with 200 mm diameter tops. Local suppliers could not supply enough of them, so some logs were shipped from Fort Good Hope at a higher delivered cost.

Our Lady of Victory Church, an igloo-shaped Catholic church, by far the largest and most impressive building in Inuvik, is not supported on piles. Its foundation consists of a gravel pad two to two and a half meters thick, which Merrill's crew provided. On this the Brothers poured a thick, reinforced concrete slab. Possibly the whole building has settled, but it has shown no signs of uneven settlement. Also the under floor framework of the church may be sufficiently rigid to bridge dips resulting from settlement. A French architect designed the building and several Brothers constructed it, all parties deserve great credit.

Figure 84. *The famous Igloo-shaped Catholic Church in Inuvik. Photo by Jack Grainge.*

Father Adam showed me the plans for the building. The complicated, arch framework supporting the dome roof fascinated me. Another remarkable feature of the church is the many, large oil paintings on the walls. The artist was Mona Thrasher, sister of Peter, whom I had met in 1954 at Mrs. Norris's trapping cabin.

In August 1956, the NWT Council held its summer meeting in East Three. Merrill chose a convenient site less than a kilometer from a construction camp at the wharf. Merrill arranged several of the ubiquitous 512 houses to provide a Council meeting place, dormitories for members and press, and a dining room.

At the 1957 summer meeting in Frobisher Bay, Knute Lang, the elected member representing the Mackenzie Delta, recommended that East Three be named Inuvik, an Inuktitut word meaning "Place of Man," or, "Where people Live." In 1958 the Government of Canada accepted their recommendation.

The top guns in DI&NA were hot to trot to get construction of the airport started in late 1955 or early 1956. Based on Merrill's description of the ample supply of dolomite rock close to the site of the airstrip, DI&NA took bold steps. First they bought and arranged for shipment of all equipment necessary for the construction — a large jaw crusher, a second stage cone crusher, an immense P & H shovel and several huge Tournapull, self-propelled earth-movers. Then they called tenders for construction before any contractors had set foot on the site.

In 1956 Aklavik Construction Co. of Edmonton, a consortium of several major contractors, including the huge Mannix Construction Co. of Calgary, met the challenge. The contractor's estimating engineers arrived during spring breakup of the Channel. Their plane could not land on the ice-choked water. Consequently they circled the site while talking to Merrill by radio. He answered their questions. He told them about the site itself, and that the dolomite in the hill near the airport could be crushed easily.

Aklavik Construction Co. submitted what they stated was a likely cost. Payment was to be the actual cost plus a fair profit. If the actual cost was less than their estimated cost, they would receive a bonus in addition to the actual cost and reasonable profit. Dynamiting the dolomite loose from the hill broke up the rock more than expected. Therefore only the jaw crusher was used, saving much time and money. Thus the actual cost was well under their estimate. They received their cost plus a profit and a generous bonus.

In a vain attempt to construct the roads as far as the airport site before the construction contractor arrived, Merrill arranged shifts of road-building crews operating throughout the twenty-four daylight hours per day. Unfortunately they ran out of thawed gravel before the road reached the airport.

In August 1956, superintendent Bill Venables began making on-site preparations for the construction of the airport. Merrill supplied him with a large warehouse and several 512 houses for use while establishing the airstrip base.

Because the road to the airport was not complete, the construction crew and equipment remained in East Three. By late October 1956, the ground had frozen and the crew moved to the airport site. They took along several 512s for offices and accommodation.

Airport construction carried on all winter long, and throughout 1957. The crushed rock layer on the runway, four meters thick, proved to be stable. Years later they added a thick layer of asphalt concrete that has also been stable. Eventually the airstrip was extended so that huge military jets could land there.

The weather station was relocated from Aklavik to the Inuvik airport. Barney O'Neill became airport manager. On 7 August 1995, the airport was named Inuvik Mike Zubko Airport. Zubko had operated the first commercial aircraft business in Aklavik and Inuvik. He continued flying until becoming incurably ill. He died in October 1991.

* * * *

In early 1957 Merrill became administrator of the Mackenzie District of the Northwest Territories. Walrath took over as project manager, under DPW administration.

In 1957 construction of major works began. Poole Construction Co. Ltd., later to become PCL, an Edmonton company with many multimillion-dollar contracts, constructed the power plant and the water and sewerage systems. Their employees' camp was situated toward Hidden Lake from Merrill's camp. They used water from Merrill's pipeline and discharged their wastewater to a shallow puddle of about forty meters in diameter. They should have set their camp so that the sewage discharged into Duck Lake. People did not walk through the wastewater puddle. It was temporary and caused no public health problem. Nobody seemed to mind, so I did not criticize it. The Company shut down work during the winter.

FENCO designed the power plant. During winter, power was to be generated by steam turbine-driven generators burning bunker C from the oil refinery at Norman Wells. During summer, when less heat was required, diesel-electric generators would be operated.

Engineers of D.R. Stanley and Associates Ltd. designed the water supply system. Hidden Lake became the summer water supply source. A pump house on stilts, located about fifty meters offshore, contained a micro-strainer for removing plankton and bugs. Although the water was uncontaminated, it was chlorinated because that treatment for water from surface sources was standard. Treated water was pumped through an insulated pipeline running along a catwalk to the shore and down the hill to the distribution system.

Government policy required plumbing services to be based on fee for service. Piped, superheated water for heating buildings, water pipes and sewers would serve government-owned buildings and others who could afford them. That drew a line between the haves and the have-nots.

Figure 85. *Water supply system using Hidden Lake. Photo by Jack Grainge.*

Figure 86. *Lemarr Hubbes showing utilidors under construction through culvert under raised road. Photo by Jack Grainge.*

Engineers of FENCO designed the superheated water pipes, and the water mains and sewers, all of which were contained in above-ground utilidors. The main utilidors were approximately 1.2 meters high and 1.0 meters wide. They were constructed of wood-frame, with insulated, aluminum-faced panels. The insulated, superheated-water pipes rested on an upper platform. The uninsulated water mains and sewers rested on the bottom. Removable side panels allowed access to all pipes. The utilidors were supported on timber piles.

Figure 87. *Utilidors causing the spread of the community. Photo by Jack Grainge.*

Figure 88. *Shallow bury water pipe and sewer constructed in 1904 at Dawson, YT, designated a landmark in 1971 by American Water Works Association. Photo by Jack Grainge.*

They took up much space and caused spreading of the community.

The pipes from the utilidors to the individual buildings were contained in utilidettes. They were similar to, but smaller than, the utilidors.

Deciding the thickness of insulation around the superheated-water pipes was too difficult a problem. Heat loss from the superheated water pipes to the household water pipes prevented the water and sewage from freezing as it was designed to do. However, in some areas, it overheated the water in the mains, especially in summer. Throughout the summers, Bide Clark, the foreman, removed some of the wall panels so that some of the hot air around the superheated water pipes would escape. In both winter and summer, residents kept jugs of drinking water in their kitchen refrigerators.

A few years after the completion of the construction of the original town of Inuvik, new residential districts were added. Piped water mains and sewers in utilidors were extended to them, as well as to some of the originally unserviced houses. To reduce construction and operational costs, the new utilidors did not contain the superheated water pipelines, and were laid on the ground surface. To prevent freezing, the water was heated and recirculated.

Figure 89. *Stop-log dam outlet to the sewage lagoon. Photo by Jack Grainge.*

Stan Copp and I had pointed out that the above-ground utilidors would break up the town. We suggested experimenting with subsurface water mains and sewers. At the time I did not know that in 1904, local, placer mining engineers constructed successful, shallow piped water and sewer systems for the town of Dawson. The town, on frost-susceptible silt, was then underlain by permafrost to a depth of nine meters. These soil conditions were far worse than those at Inuvik.

Figure 90. *Aerial view of Dawson. Photo by Jack Grainge.*

* * * *

When I visited the settlement in 1956, I was surprised to see in the un-serviced area two, extraordinary, two meters by two meters, two-story, service buildings. They contained water and sewage tanks, with pumps. A water tap was at the lower level. To empty a bucket toilet, a person had to climb outside stairs to a second-story chute. Probably the water from these buildings was used, but I doubt that anyone carried a toilet bucket up the stairs. Later the Town began a water and sewage haulage service. The operators of the garbage truck hauled away the toilet bags. They also demolished the service buildings.

The Northern Canada Power Commission operated the water supply, sewerage, and superheated-water heating systems. Bide Clark, a short, broad-shouldered fellow, kept them operating well. He used the two sources of water, Hidden Lake in summer and East Channel in winter, when the water was clear. During the winter, surplus water being pumped from the river replenished the water in Hidden Lake. It was many years before the increased water demand in summer made it necessary to pipe water from a lake five kilometers away.

Figure 91. *View from high hill overlooking Dawson, YT. Photo by Jack Grainge.*

Clark had one hang-up. When fluoridation of the water supply was ordered, he did not agree with the policy. Therefore he ignored instructions to operate the fluoridator. It did not surprise me. At that time many critics, including a few doctors and dentists, were claiming fluoride to be a dangerous poison. To be safe its concentration in drinking water must be carefully controlled. However Clark did not ignore the second order to operate the fluoridator.

Clark was quick to accept another of my suggestions. Each spring he had been melting the ice in a road culvert with a steam jet. I suggested that he lay a heating cable through the culvert and in spring run an electric current through it. That melted a small hole and started a trickle through the culvert. Within a day the trickle grew to a substantial flow.

* * * *

I had suggested an unusual design for the sewage lagoon. It was to use an existing long narrow pond alongside the channel. Because there was water in the pond throughout the summer, I assumed that the embankment around the lagoon was relatively water tight. I recommended a stop-log dam at the far end to increase the lagoon's capacity so that it could hold all of the year's sewage. In the spring the stop logs would be removed to allow the putrid, undecomposed sewage to be washed away by the spring flood.

During the summer, the twenty-four hours-a-day of sunlight would stimulate the growth of algae. This would oxidize the fresh, sewage in the lagoon and render it reasonably odor free. Therefore the lagoon could be closer to the community than would ordinarily be acceptable.

Figure 92. *Victor Georgevick, one of six scientists from Russia that I showed around Edmonton and Inuvik. Photo by Jack Grainge.*

Engineers of D.R. Stanley and Associates Ltd., Edmonton, agreed with my general suggestions regarding the sewage treatment system. However they did not consult me regarding its design. The inlet sewer ran through a two-meter by two-meter, wood-frame hut on stilts, a meter above the level of the lagoon, and approximately twenty meters from the shore. The hut was to be a place to install a sewage heater if the sewage was too cold to penetrate the ice on the lagoon.

To spread the sewage sludge, Stanley's engineers designed unusual outlets discharging to the lagoon. Leading further into the lagoon from the hut was a ten meter long catwalk. Two ten-meter-long pipes from the end of the sewer in the hut led to points, one on each side of the catwalk. Stanley's engineers had intended that the end of the each pipe should be connected by ropes extending, one to the catwalk and the other to the shore of the lagoon. From time to time men on the catwalk and shores would pull the ropes to move the ends of the pipe outlets to different places.

Unfortunately none of us foresaw the formation of permafrost lifting the embankment above the floodwater level. Consequently the sewage sludge did not get washed away during the spring flood. Furthermore, seepage from the lagoon increased, probably due to the rise in the height of the lagoon above the channel. The lagoon was shallower than expected. It did not even cover the sewage sludge, at the sewage discharge points. These pipes became trapped in the sludge. Unfortunately too the town grew closer to the lagoon than originally planned.

At the same time, silt was being deposited on the nearby shore of the channel. Permafrost uplifted this emerging deposit and resulted in the formation of a small, nature-made, airplane runway. It was similar to the airstrip that had developed at Aklavik. Airplane passengers had to pass by the stinking sludge at the near end of the lagoon.

During the late 1960s, John Parker, Deputy Commissioner of the NWT, called a public meeting in Inuvik to discuss what should be done about the odors. I believe I suggested deepening the lagoon at the inlet end. This might have made some improvement. I am sorry that I did not think to suggest extending the sewage outfall point farther into the lagoon. Then the malodorous sewage sludge would be immersed in liquid and also be farther away from the road to the airport.

Building another lagoon farther distant was out of the question. The Canadian Navy operations buildings and the IOL tanks were beyond the lagoon. Therefore a new lagoon would require a costly sewage pump station and more than two kilometers of heated and insulated pipe to reach a suitable lagoon location.

The town council hired Associated Engineering Services Ltd. to suggest a solution to the problem. Norm Lawrence, the president, recommended pumping the sewage through a pipe buried in the channel bed to discharge into a large lake on the nearby delta island. People objected because they used that lake for trapping and recreation.

Probably few residents of Inuvik thought that a sewage lagoon was a satisfactory sewage treatment facility. During an outbreak of infectious hepatitis, an intestinal disease, among children in one of the school dormitories, someone showed an investigating epidemiologist the lagoon. The guide might have inferred that it was not a satisfactory treatment system. In a forgetful moment the epidemiologist concluded that the lagoon was the cause of the outbreak of that disease. He so reported.

Outbreaks of infectious hepatitis are common in institutions, particularly among children, who are generally negligent in hand washing. The outbreak of hepatitis among the school children in the dormitory was typical.

I phoned the epidemiologist in Toronto to ask him why he thought that the sewage lagoon was responsible for the spread of that disease. He replied that he was told that the sewage lagoon provided substandard treatment. I explained to him that lagoons were the most common sewage treatment system in the prairie provinces and states. Also Bob Dawson, an engineer of our office, had made tests that showed that the lagoon was providing good treatment. The epidemiologist apologized.

John Banksland, a likable, young Inuk originally from Sachs Harbour, who had taken a course in sanitation practices at Ryerson Polytechnic Institute in Toronto, helped in containing the outbreak. Banksland had attended the Catholic residential school in Aklavik. While there he had been a boy scout and had attended a Boy Scout jamboree in London, England. He deserves great credit for taking the Ryerson course long after he had graduated from high school. I encouraged him to do so. He and his wife and babies visited us while he was apprenticing in Edmonton, partly under my supervision.

Figure 93. *John Banksland and daughter. Photo by Jack Grainge.*

References:

Lawrence, N. 1996. Personal communication
Merril, C. 1996. Personal communication.
Schaeffer, O. 1996. Personal communication.
Zubko, D. 1996. Personal communication.

FORT McPHERSON

Figure 94. *Aerial photo of Fort McPherson. Department of Energy, Mines, and Resources, Canada, 1979.*

Fort McPherson is situated on a high hill on the right (east) bank of the Peel River, fifty kilometers above its confluence with a channel of the Mackenzie River delta. The land occupied by the community slopes gently southeast toward a string of lakes parallel to the river shore.

The grassed topsoil overlies a thin layer of muskeg, below which is a shale bedrock. Permafrost is in the ground, probably to depths of forty, fifty or more meters. However the ground is stable. It does not heave or settle unevenly due to thawing of ice in the ground. I presume the ground does not contain ice layers.

After Sir John Franklin had passed through the area on his second expedition (1825-28) he advised the HBC of the fur riches in the area. Soon thereafter the North West Trading Co., from Alaska, established a post there. To haul their merchandise and equipment up the high river bank, they installed a machine-operated belt system.

In 1840 Isbister and Bell established a HBC post there. Eight years later they named it after Chief Factor Murdoch McPherson. In 1852 Dene in a nearby camp moved to that high location from which they could see a long distance north. Thus they would be safe from surprise by Inuit bands, with whom sometimes they had been feuding.

Figure 95. *Aerial view of Fort McPherson, 1974. Photo by Ev Carefoot.*

In 1872 John Firth came to the Mackenzie River Region of the HBC. Eleven years later he became factor at Fort McPherson. One of his four children, William, taught himself to play the fiddle. He could hear a tune, hum it over in his mind a few times, and then play it. William's son, Wally, became a member of parliament. During the 1960s and 1970s, Shirley and Sharon Firth, great granddaughters of John Firth, won prizes in international cross-country skiing races.

A contemporary of John Firth was Sara Simon, the many-talented Dene wife of the Anglican minister. She served as midwife throughout the area and played hymns on her portable organ. When Indian Agent Garry Hunter arrived on his annual treaty-paying visits, she provided him with details of all the births, deaths and whereabouts of everybody in the camps throughout the area. Therefore he did not need to make the rounds himself.

In 1903, the North-West Mounted Police, later to become the Royal Canadian Mounted Police, established a post on Herschel Island, off the north coast of the Yukon Territory. They collected license fees from American whalers and thereby established Canadian sovereignty. They built another post at Fort McPherson with Sergeant (later Inspector) Francis J. (Frank) Fitzgerald in charge of both posts. At the time Fort McPherson consisted of five HBC buildings, an Anglican church and house, a ramshackle building rented by the RCMP and three log cabins owned by Dene.

Figure 96. *Main Street Fort McPherson, 1958. Photo by Jack Grainge.*

In December 1910, Sergeant Fitzgerald and three constables began the first of annual winter patrols by dog teams from Fort McPherson to Dawson. Unfortunately they did not hire a Dene guide. Short of half way, they failed to find the right route. They attempted to return to Fort McPherson, but they starved only fifty kilometers from home.

In 1949, Nurse Dawn Smith came to Fort McPherson. Later that year she went to Aklavik to marry Mike Zubko. Inspector Lewis Watson, RCMP, officiated.

* * * *

I first went to Fort McPherson in 1958. The community consisted of the HBC buildings, a store operated by Mike Krutko, a four-room school, a students' dormitory, a teacherage, a DI&NA office and house, a diesel-electric generator in a building, a nursing station, a DOT communications-and-weather station, an RCMP post, Anglican and Catholic missions and a few residences.

People carried water from a ten-hectare lake, located two hundred meters southeast of the community. That lake overflowed northward into a three-hectare lake east of the settlement into which sewage and run-off water from the town

Figure 97. *River bank erosion. Photo by Jack Grainge.*

flowed. The two lakes were part of a chain of lakes that flowed northward, parallel to the Peel River. During annual spring floods, the Peel River overflowed into a lake south of the community and flowed through the chain of lakes, discharging back into the Peel River from a lake downstream of the community.

After discussions with territorial engineers and local people, we recommended that the water be piped from the drinking-water lake to the buildings containing running water systems. We added that the sewage from them be piped to the lower lake. We realized that there are sound reasons both for and against accepting these recommendations. The high cost of constructing and maintaining these difficult systems must be balanced against the inevitable contamination of the hauled water and spills from household sewage holding tanks. Also when the community grows, the cost of piping will become less costly than haulage.

Figure 98. *Dene resident cooking outside, as it was too hot inside. Photo by Jack Grainge.*

In approximately 1961, Herb McNabb, an enthusiastic, dark, fine-featured young engineer of the federal DPW came to Fort McPherson. He planned and constructed above-ground, water supply and sewer systems for buildings with plumbing. The land, sloping toward the sewage lake, favored this construction.

A peninsula and a couple of islands separated the two lakes, but still there was a narrow connection between them. Mr. and Mrs. Keith Billington, nurses, assured McNabb that the flow of water was always northward, from the water lake to the sewage lake. Therefore the sewage would not pollute the lake water.

Regardless of Billington's assurances regarding the safeness from sewage pollution of the lake water, some people claimed that the two lakes should be separated by a dam. McNabb hauled earth to dam the connection between the two lakes. He soon found that the job was beyond his budget, and gave up. Later federal engineers completed the dam.

To stay within budget, McNabb did not connect the nursing station to the public sewer system. Sometime previously, the DPW had constructed a septic tank and seepage pit for the nursing station. However the soil around the seepage pit proved to be impermeable. Mike Pich of our office constructed a small sewage lagoon east of the building. A pump intermittently discharged the effluent in the seepage pit through a 25 mm, surface pipeline, covered with a meter of moss, to the lagoon. Before pumping the septic tank effluent, the pipe was heated with a Pyrotenax, electrical heating cable.

Figure 99. *Water intake, 1972. Photo by Jack Grainge.*

With careful attention by Billington, the system worked for a year. The following year DI&NA built a gravel roadway across the pipeline. That winter the pipeline froze. We never managed to make the system work again. We thought of shortening the distance to the lagoon, but decided against doing so. Later the building was connected by a utilidor to the public sewer system.

Pich also helped Mike Krutko construct a similar sewer system. The pipeline was buried but ended before the road that had stymied the nursing-station system. Krutko's system worked well.

The federal engineers installed a complicated, manufactured water treatment plant. I visited the community a year or two later. The plant was difficult for an operator in a remote settlement to operate. The Caucasian operator managed, but with great difficulty.

In the 1960s and '70s, DI&NA provided the Dene with wood-frame houses. They continued to operate the piped water distribution system, and trucked water to houses that were not connected to it. I am pleased to think that my ideas and friendship with enthusiastic, energetic Herb McNabb had resulted in a large part of the community being served with a piped water distribution system.

References:

Hazen, L. 1998. (Great grandson of John Firth and son-in-law of Dawn Zubko). Personal communication.

Lawrence, N. 1998. Personal communication.

Schaefer, O. 1998. Personal communication.

Zubko, D. 1998. Personal communication.

FORT GOOD HOPE

Figure 100. *Our Lady of Good Hope Church. Photo by Jack Grainge.*

Figure 101. *Stations of the cross. Photo by Ev Carefoot.*

In 1805 the North West Co. established Fort Good Hope. At thirty kilometers south of the Arctic Circle, it was for many years their most northerly trading post. In 1859 Father Grollier and some Dene moved there from Ft. McPherson. Six years later Father Emile Petitot arrived and built the present Our Lady of Good Hope Church. In 1878 Petitot began painting spectacular, red-predominating, murals, to grace its walls. Although it is twenty-eight years since I entered the front door in 1969, I can still see, in my mind's eye, that beautiful scene.

In 1969, Mike Pich of our office tried to take a picture of the murals surrounding the alter. However all views were spoiled by an ugly cable and turnbuckle. It ran from one sidewall to the other to prevent them from bulging outward. This is a common fault of buildings without wall-height ceilings. The joists in these ceilings hold the walls in place. Most likely they would have later moved the turnbuckle to the tops of the walls.

The community is situated on the right bank of the Mackenzie River, on wedge-shaped land between Jackfish Creek and the river. The land slopes southward toward the creek. The wharf fronts the river at the bottom end. The community is near the border between the zones of continuous and discontinuous permafrost. The community faces south, so no permafrost exists in the soil.

Approaching by plane from the south in 1956, I found the community to consist of attractive houses, mainly white, glistening in the noon sun. Grass surrounded the houses. Some of the people had planted borders of flowers near their front doors.

Two draws divide the community into three high areas. The mission, RCMP, nursing station, Department of Indian Affairs and Northern Development administrator, game warden and HBC occupy the southern-most hill, the Department of Transport the middle one, and a school the northern one.

Figure 102. *Aerial view of Fort Good Hope, 1969. Photo by Ev Carefoot.*

In summer the people set a portable Wajax pump in Jackfish Creek, pumping water through a hose to the water tanks of those households with pressure systems. The people preferred the colored water from Jackfish Creek to the turbid river water. Neither Pich nor I recommended chlorinating the water, because the chlorine would have reacted with the organics in the water and produced an unacceptable taste.

After ice formed on the river in winter, the water clarifies. People then drew water or cut ice from the river. Dog teams hauled both the water and the ice.

* * * *

Honey buckets were common. The nursing station and a DOT house contained running-water plumbing with sewage disposal to septic tanks and seepage pits. However seepage was negligible. I suggested that for a trial they pipe the septic tank effluent to a small excavation part way down the river bank. The effluent seeped away through the bank. For a trial in the first winter, the nursing station effluent was diverted to bypass the lagoon but the DOT sewage did not do so.

During winter, ice prevented the sewage from reaching the river. During spring breakup, the flood water would rise and flush away the sewage on the bank at a time when the river was so muddy no one would take water from it.

When the ice on the DOT pond thawed in the spring, the odor became insufferable. The tenant endured that odor until I returned. I concluded that a household-size lagoon receiving septic tank effluent needed to be much larger than I had previously thought. I recommended that in future the sewage from that building should bypass the pond during winter.

The DOT followed up their success by constructing a large, one-cell lagoon thirty meters north to serve all of their houses. The effluent pipes were contained in insulated boxing together with Pyrotenax, electrical heating cables. During winter, the operator turned on the heating cable before pumping the effluent from the septic tanks. He drained the lagoon during the spring breakup of the river and in the late fall.

In time all government-owned buildings and the HBC had septic-tanks. When I was there in 1969, both the RCMP and the game warden discharged their septic tank effluents over the river bank. It was downstream of the wharf so the water systems of the vessels docking there were unaffected. However for the sake of people who might live downstream, the public health benefit of a sewer system for the many residents in the community outweighed the danger to people who might camp a few kilometers downstream. Besides, the dilution was tremendous and the people downstream could be warned of the problem.

The effluent from the sewage system of the two-room school and teacherage, a one-room school, and two houses was piped sixty meters to underground seepage pits near the river bank. There was no downhill surfacing of effluent. Over the years, several of their pits had failed and been replaced.

The Dene community health worker had been encouraging his people to dispose of toilet and wet wastes to wood-walled, individual, household seepage pits with covers. He constructed them for widows and older people. The pits worked well. He was an imaginative, effective, community health worker.

* * * *

In 1962, engineers of Strong, Lamb and Nelson Ltd. examined the community. They estimated the costs of various alternatives for piping water throughout the settlement. It was prohibitive. No improvements resulted from their study.

In 1969 the summer source of water was a large, partly spring-fed pond, located 1.5 km north of the community. In winter the contractor pumped water from the river to the pond. The water pumped from the pond was somewhat colored.

Corless Construction Co. delivered water to household steel tanks in the government-owned buildings and to four 4500 liter community tanks among the Dene residences. Mr. Corless added chlorine solution at the time of filling the haulage tank. After about twenty minutes chlorine contact time, he tested for its adequacy. He charged two cents per gallon for the water. For drinking water during the winter, the Dene hauled ice from the river, and occasionally gathered snow.

In his report, Lawrence recommended long range planning so that eventually a pipe distribution system would be feasible. Danny Makale, a town planner, prepared such a plan, but the cost of the construction was prohibitive.

The contractor had been maintaining a ditch-and-cover garbage disposal site. Natural drainage led to Jackfish Creek. Lawrence recommended moving the disposal site to a remote location near the river, a half kilometer north of the water supply lake.

References

Carefoot, E. 1998. Personal communication.

Grainge, J. 1970. Reports on Fort Good Hope, examination dates October 1969 and October 1970. Reports in the possession of Ev Carefoot, formerly an engineer in Associated Engineering Services Ltd.

Lawrence, N. 1970. Reports on Fort Good Hope, examination dates August 1969 and March 1970.

Lawrence, N. 1998. Personal communication.

DELINE (FORT FRANKLIN)

Deline is situated on the southwest end of Great Bear Lake, six kilometers from the right bank of Great Bear River. It extends almost a kilometer along the shore. In 1961, I arrived there on the weekly mail flight from Norman Wells. As seen from the air that sunny day, it was a picturesque community, a row of buildings parallel to the shore, with a fine beach in front, surrounded by a scattering of conifers and bushes, which were interspersed with areas of mosses, lichens and grass. From east to west there were a three-room school, a teacherage, a Catholic Mission, several log cabins and the white, red-roofed, HBC buildings.

Originally a few Dene lived near the mouth of the river where fish were bountiful. In approximately 1819 the North West Co. established a trading post there and later abandoned it. Through1825-27 the HBC maintained the post for Sir John Franklin, who occupied it as his winter headquarters. He was mapping the Coppermine River and the mainland coast east of Coppermine.

Figure 103. *Aerial view–approaching Fort Franklin, 1969. Photo by Jack Grainge.*

In 1949-50 an HBC post, a federal day school and a Roman Catholic mission were established there. Artistic Father Rene Fumoleau built the eye-catching, beautiful St. Teresa of Avila Church. It was white with a red tipi-shape roof. After it burned down, Brother Maurice Larocque rebuilt it. Father Fumoleau took picturesque color photos, some of which brighten his book, *Denendeh.*

In 1930 Gilbert LaBine discovered a deposit of silver, radium and the then valueless uranium at Port Radium, on Echo Bay, on the east shore of Great Bear Lake. White Eagle Mines, using wood-burning, steam-engine-driven vessels supplied LaBine's Eldorado Mine. They sailed up the 120-kilometer-long Great Bear River and across the lake. Many Dene migrated to Fort Franklin to work for the company. They gathered firewood and transferred freight to and from 75-cm draft vessels used to navigate rapids half way down the river.

In either 1935 or 1936, LaBine bought Northern Transportation Co. from White Eagle Silver Mines. He also bought shallow-draft vessels and barges for freighting over the eleven kilometers of rapids on the Great Bear River. There was also a road and an oil pipeline around the rapids. Dene in both towns handled the freight for the Company.

The miners dynamited, sorted and bagged the radium ore and everyone—miners, packers and shippers—handled the sacks with no knowledge of any possible danger to their health. The Dene at Deline transferred the bags of ore, the ore containing uranium and radium, to shallow-draft barges at the rapids on the Great Bear River and then back to deep-draft barges on the other side of the rapids. Later they portaged the freight around the rapids.

In 1942 scientists in USA needed uranium with which to produce atomic bombs. The Canadian government nationalized the mining company, and the Northern Transportation Co. Ltd (NTCL) and established an airline. The mining company, Eldorado Mining and Refining Ltd., milled the ore to produce uranium oxide, which they bagged and shipped to their refinery in Port Hope, Ontario.

Some of the uranium oxide was shipped to Edmonton in spare space on company planes. Most of it was shipped by barge to Fort Smith and by truck across the portage to the railhead at Waterways. The Dene at Deline supplied the vessels with firewood and also transshipped the bags of uranium oxide. In doing so the workers would have received low levels of radon exposure.

The general population is exposed to some radiation exposure. New cells replace the radon damaged human cells. Studies of atomic bomb survivors who were exposed to moderate radiation had lower death rates, than people not so exposed. Studies have shown that no apparent harm results from exposures up to about five hundred times background levels (Harris 1997, Konde 1993).

* * * *

The people in the community of Deline carried water from the lakeshore. A contractor hauled water for the school and teacherage. Others bucketed water from the lake. Water taken near the shore contained runoff water from the shore, which was slightly colored. The water in the lake is clear and soft. During winter, water was obtained through holes in the ice and by melting blocks of ice in barrels in the houses.

People threw their wash water on the ground near their doorways where in summer it seeped away through the silt, sand and clay soil. Melting snow and drains washed some wastes into the lake. Many people had pit privies. Others emptied honey buckets further inland.

* * * *

In 1962 when both Mr. Cottrell, the region administrator of Indian Affairs Branch, and I happened to meet in Yellowknife, he asked me for advice. I suggested a summer water distribution system buried about three or four centimeters below the surface of the ground. Because of the sloping ground in Fort Franklin, the pipe could be drained in the fall and left in place throughout the winter. I offered to provide Tom Brown, a new engineer on our staff, as supervisor for the project. Cottrell would supply the materials and two local laborers.

They constructed a system consisting of a centrifugal pump in a four-meter-by-two-meter wooden shed on the shore, a 75-meter long, fifty-millimeter, plastic suction pipe, a 135-liter pressure tank and six hundred meters of 38-millimeter plastic distribution pipe.

Figure 104. *Dye study. Photo by Jack Grainge.*

The suction pipe was weighted down with widely spaced rock-filled oil barrels. An empty, floating oil barrel served as a buoy to show the location of its end. The end of the pipe was fitted with a fine wire screen and suspended 60 cm above the lake bottom in 2.1 m depth of water.

The distribution pipe led to the government-owned buildings and six standpipes distributed among the houses without indoor plumbing. Gate valves and drains were placed at three low spots in the line. Brown explained that the system should be drained in the fall, and left in place throughout the winter. This could be accomplished by disconnecting and draining the equipment in the pumphouse and opening the taps at the hydrants and the gate valves at the three low points.

That fall we received a letter from a teacher at Fort Franklin stating that he had taken up all of the pipes and piled them safely. I suppose the instructions to leave the distribution system had not been passed along to him. So the next summer John Shaw, an engineer in our office, went there and reinstalled the system. Because the plastic water intake pipe was difficult to sink below the water surface, Shaw replaced it with galvanized iron pipes.

In 1965, DNH&W built a one-nurse, nursing station near the teacherage and school. The buildings being close together made it possible for one set of water and sewer systems to serve all three buildings.

To avoid polluting everybody's drinking water, they dug a lagoon. It was located near the shore, about a hundred meters northwest of the school. The lagoon was small, about sixty meters by forty meters, but large enough to handle septic tank effluent from the three-room school, teacherage and nursing station.

In order to be sure of a water intake point that was safe from pollution , we studied the currents in the lake. In July 1965, Bob Dawson and Mike Pich, engineers in our office, put Rhodamine B, a harmless red dye, in the lake a short distance off shore in front of the wharf. Depending on wind directions the water flowed parallel to the shore, at different times in opposite directions, but always gradually diverging from the shore. Dye that was set out on July 14 was blown almost three kilometers east by a northwest wind, and was returned the next day, widely spread, two hundred meters offshore by a northeast wind. We concluded that a water intake 150 meters offshore at the proposed new wharf would be safe. If the water were safely chlorinated it could be taken much closer to shore.

Late in the winter, John Shaw returned to study currents in the lake under ice cover. He found that the dye placed at holes off shore at the wharf diffused slowly in all directions. Our conclusion, the current was negligible.

In 1969 I returned to Fort Franklin. In the previous five years the population of the community had grown from 270 to 410. Materials were on hand for Department of Indian and Northern Affairs (DI&NA) engineers to construct a new water intake and pump at the wharf. It would pump water directly to an 80,000-liter tank in a heated and insulated building near the school.

A contractor with a tractor-drawn trailer delivered water to the water tanks in houses with pressure plumbing systems. As so often happened with such equipment, the delivery hose nozzle was subject to contamination by being dragged on the ground and then strewn haphazardly on the trailer. I told the contractor that he should wind the hoses on reels so that they would not drag on the ground.

The contractor, using a tank on a trailer, hauled the sewage from the buildings containing holding tanks to a new sewage lagoon. Sewage was overflowing from one side of the lagoon toward Airplane Lake, on which planes occasionally landed. People objected because the lake was a popular fishing place. Also, often small planes landed there rather than on Great Bear Lake. I note that in 1977, long after I was involved, a large, deep lagoon was constructed.

A community plan showed a future road to be constructed between the nursing station and the school and teachers' residence. In my report I pointed out that there seemed to be no reason for that road route. Such a road would separate two buildings, which logically should be connected to the same water and sewer facilities.

I have been disappointed that we were unable by our work to convince the planners how simple it would have been to construct buried water mains and sewer systems in Deline. We had demonstrated that the soil conditions were suitable for the construction of sanitary subsurface water mains and sewer systems and a sewage lagoon, at least for summer use. It shows the difficulties that result from there being a high rate of turnover of engineers in the planning offices.

References:

Government of the Northwest Territories. 1981. Water and Sanitation Manual and Pamphlets. Yellowknife: GNWT.

Harris, W.E. 1997. *Low Dose Risk Assessment.* Edmonton: University of Alberta

Konde, S. 1993. *Health Effects of Low-Level Radiation.* Osaka, Japan: Kinku University Press.

Pich, M. 1968, 1998. Personal communication.

Shaw, J. 1968, 1998. Personal communication.

NORMAN WELLS

Norman Wells is situated on the right (north) bank of the Mackenzie River, 130 kilometers south of the Arctic Circle. In 1911 J.K. Cornwall examined petroleum oil seepages from the banks of Bosworth Creek, and nearby banks of the Mackenzie River, and three years later P.O. Bosworth staked three oil claims. In 1918 the Northwest Co. bought his claims. In the following year, they drilled Discovery Well and built a small refinery. In 1939 they brought in three more wells. They rebuilt the refinery, producing a wide variety of products. Three years later, the US Army began constructing the Canol oil pipeline to Whitehorse, an airstrip in Norman Wells and an addition to the refinery. They also drilled several new oil wells.

Figure 105. *Aerial photo of Norman Wells area, showing oil wells, refinery, and Canol pipeline 'junk heap' (Department of Energy, Mines, and Resources, 1961). The town of Norman Wells demonstrates US Army influence in Canada.*

In 1953, I returned from Aklavik with Stewart Hill flying a single engine, Norseman airplane, known then as the workhorse of the North. Mr. Hill, he was not a first-name kind of person, docked at the wooden wharf. We passengers carried our baggage up the fourteen-meter bank to the CPA guest house. I stayed there three days awaiting the weekly DC3 plane to Edmonton. Meanwhile Mr. Hill flew passengers and freight to and from communities in the surrounding district.

The community had been growing on Imperial Oil Ltd. (IOL) property between Bosworth Creek and the refinery to the west. It consisted of an RCMP post, a hospital, a church for all faiths, a recreation hall and men's quarters. West of the refinery were houses of married employees. East of Bosworth Creek were the CPA guest house and I think one or two other buildings. A three kilometer long road on firm ground alongside the high river bank led to the DOT weather station, a Canadian National telecommunication station and CPA airport building at the southeast end of the airstrip. The airstrip built on a gravel esker, diverged inland from the river.

The CPA guest house was a simple, one-story, wood-frame building without a basement. Vern and Thelda Olsen, from Cardston, Alberta, and a young wife of an IOL employee kept it clean,

Figure 106. *Oil storage tanks, many built by the US Army. Many of the riveted plates leaked. Photo by Jack Grainge.*

Figure 107. *Perpetual oil seepage burning on a bank of the Mackenzie River. Photo by Jack Grainge.*

Figure 108. *Early settlement of Norman Wells. Photo by Jack Grainge.*

and served tasty, family-style meals. I vividly remember one evening with the friendly Olsens and the jolly airline crew and guests. The CPA hostess related hilarious tales about some of the passengers she had met and about spilling hot soup on a bald man's head. We played blackjack. I lost. Years later Olsen served me a free dinner at his restaurant in Cardston, Alberta.

Olsen cultivated a small vegetable garden on a southern slope at the top of the river bank. He mixed sawdust with the sandy soil. With the aid of fertilizer and long summer days, his garden grew quickly. Following the same formula, the refinery superintendent's wife cultivated a beautiful flower garden around their house.

By coincidence my young half-brother, Rupert Littke, a university student, had a summer job working in the IOL truck maintenance shop. He took time off to show me a huge dump of abandoned equipment left by the American Army at the furthermost end of the airstrip. The community continued to use equipment from that dump. On June 21, Rupert and several IOL employees had climbed a nearby hill to see the midnight sun. As is often the case for that date, clouds hid the sun.

* * * *

The IOL water supply source was a pond created by a small dam on Bosworth Creek. The creek flows approximately fifty kilometers through a relatively uncontaminated area from the Franklin Mountains.

The water is very hard, somewhat colored and subject to minor contamination, but otherwise satisfactory. Late in some winters, the creek stops flowing, and the water becomes extremely hard. Some winters it was necessary to pump water from the Mackenzie River into the pond. In the fall, when ice forms on the river, the water becomes clear and free of pollution. It is slightly hard.

The water distribution pipes, sewers and the steam pipes for heating buildings were contained in insulated, wood utilidors, supported on piles. However some of the piles had shifted out of line. Permafrost in ice-rich soil below them had melted. John Pihlainen, Hank Johnson and Roger Brown, engineers of the National Research Council in Ottawa, had from 1952 to 1953, conducted permafrost research nearby. They found horizontal ice layers in the permafrost ground (temperature permanently lower than 0°C). The oil well drillers reported that the permafrost extended to a depth of three hundred meters.

When I was there in 1956, the Company had rebuilt the utilidors, laying some of them on piles and others on mounded earth, the type depending on the ice content of the subsoil. The new utilidors lasted several years with little maintenance.

Two or three years later, IOL raised the elevation of the dam, but the community grew and during some winters they again ran out of water. In 1969, Norm Lawrence of Associated Engineering Services Ltd. (AESL) recommended a dam a kilometer upstream. Its construction ended the water shortages. IOL installed a hypochlorinator and chlorine contact tank.

* * * *

Effluent from several septic tanks, discharged over the high river bank. In summer these small streams, which contained no sewage solids, flowed across the shore and into the river. During winter the effluent freezes on the river bank. The spring flood, eight or more meters high, washes away the accumulation at a time when no one uses the muddy water in the river. In my 1954 report, I did not criticize that sewage effluent disposal. To have suggested pipelines down the bank to the river would then have been impractical, because ice jams during the spring floods would have ripped them away.

Stan Copp, my former boss, and I had been criticizing sewage from communities entering the river, because until the 1960s Dene lived nearby downsteam of most communities and used the water. The septic tank effluent flow was miniscule compared to the huge, fast-flowing Mackenzie River, and there were no Dene living immediately downstream. Obviously the pollution was within dilution standards.

In 1968 the Canadian Wildlife officers in Edmonton, who drank coffee with us in our Edmonton laboratory, told us about oil leaks to the river. As a result, a hundred or more migratory birds had died. Mr. D. Lapp, Game Warden, Fort Good Hope, told me that he had seen oil slicks up to fifty kilometers long downstream of the refinery. He said that they were entirely separate from the small natural oil seepages upstream from Norman Wells. He added that since the oil slicks had become larger, whitefish, herring and inconnu had disappeared from that part of the river.

Figure 109. *Oil seeping into the river, 1965. Photo by Jack Grainge.*

* * * *

Engineers from our office, John Slupski, Bob Dawson and I, examined both the water supply and the discharges from the community. IOL was planning to drill many more oil wells and increase the production rate of its refinery to fifteen times its present rate.

We found that the oil separator was too small. Oily water was draining to the river flatland where the oil was burned. However some overflowed to the river. In winter the dense black smoke from the burning puddle of oil was settling on the snow. This blackened the white ptarmigan, making them easy prey for foxes.

We recommended additional oil separation equipment. Also leaking, bolted, steel tanks should be welded tight, and embankments around the oil tanks should be repaired.

* * * *

The 240 population was expected to skyrocket. IOL. was planning to build many houses east of Bosworth Creek, which had become the main community. That section had grown and spread out along the road almost half way to the airport.

A hotel and Department of Transport (DOT) and Northern Transportation Co. Ltd. (NTCL) offices and houses were grouped at the airport-end of this road. Sewage from these buildings flowed, some to septic tanks and others to small ponds, all of which spilled over the river bank, either through pipes or by way of ditches. In a couple of cases there was no treatment. There were ten separate disposal points. A few Dene, looking for work, had moved onto the river flats nearby downstream.

Figure 110. *Small sewage pond provided some aeration treatment. Photo (1962) by Jack Grainge.*

The Department of Environment (DOE), which had just been formed, instituted a policy requiring secondary treatment (solids removal plus oxidation) of all sewage discharged to lakes and streams. We realized that small aeration systems are difficult to operate and would not destroy bacteria. The purpose of oxidizing sewage is to reduce the consumption of oxygen dissolved in the receiving stream. However the Mackenzie River is immense so that the dissolved oxygen effect of the small sewage flows from Norman Wells would be negligible. We recommended piping the sewage to some of the sloughs on the opposite side of the airport, in which the human bacteria would soon die. However the Department of Transport (DOT) objected, saying that sewage-eating bugs would attract birds which would pose a danger to airplanes.

In about 1975 the Government of the Northwest Territories (GNWT) municipal engineers prepared plans for a new subdivision east of Bosworth Creek. I suggested to the director and senior engineer that they should make sure that the plan took advantage of the topography so that a gravity sewer system could be constructed. I recommended the sewage be piped to a slough the other side of the airport. They were not impressed.

After I retired, the Water Board hired Dr. Dan Smith, a university professor, to assess the health problem. He put dye in the river and tested for it downstream. He concluded that dilution was sufficient to render the downstream water safe for drinking. Dr. Schaefer, who was serving on the NWT Water Board, asked me for my opinion. I replied that because the town was growing, they should pump the sewage to a lagoon. Eventually the community grew much larger than I think Dr. Smith realized. The NWT Water Board required a long retention secondary sewage treatment lagoon.

After I retired, Warren Suitor, a former IOL engineer told me how the DOE found fault with the appearance of abandoned, empty oil barrels on the shores of

Figure 111, 112, & 113. *Typical pile-ups at upstream bends of the Mackenzie River. Photos by Jack Grainge.*

remote lakes and rivers. Imperial Oil Ltd. assigned him, with a small crew and bulldozer, to crush and bury them. He said it had been a great way for him to spend his last two summers before retirement. Those barrels have useful purposes and eventually people in remote places will wish some of them had been spared.

References:

1968 and 1969 reports by John Shaw, John Slupsky and Jack Grainge.

1969 report by Norm Lawrence, Associated Engineering Services Ltd.

Ev Carefoot, Associated Engineering Services Ltd., had saved all three of these reports in his private files at his home.

TULITA (FORT NORMAN)

Tulita is situated on the right bank of the Mackenzie River, a kilometer south of its confluence with Great Bear River. In 1810 the North West Co. established a fur trading post there. After the union of the two trading rivals, the HBC moved their post a few times. In 1872 they returned it to its original location. It served the Dene traveling along the Great Bear and Mackenzie Rivers and the Gravol River on the opposite side of the Mackenzie River. Tulita means, "Where Two Rivers Meet."

In 1895 Charles Camsell, at age nineteen, began two years of work for the Anglican Mission. At that time, the settlement consisted of the HBC post, with stockade, and Catholic and Anglican Missions. The total population was twenty, eight of them children, Camsell's students. Most of the students were outsiders.

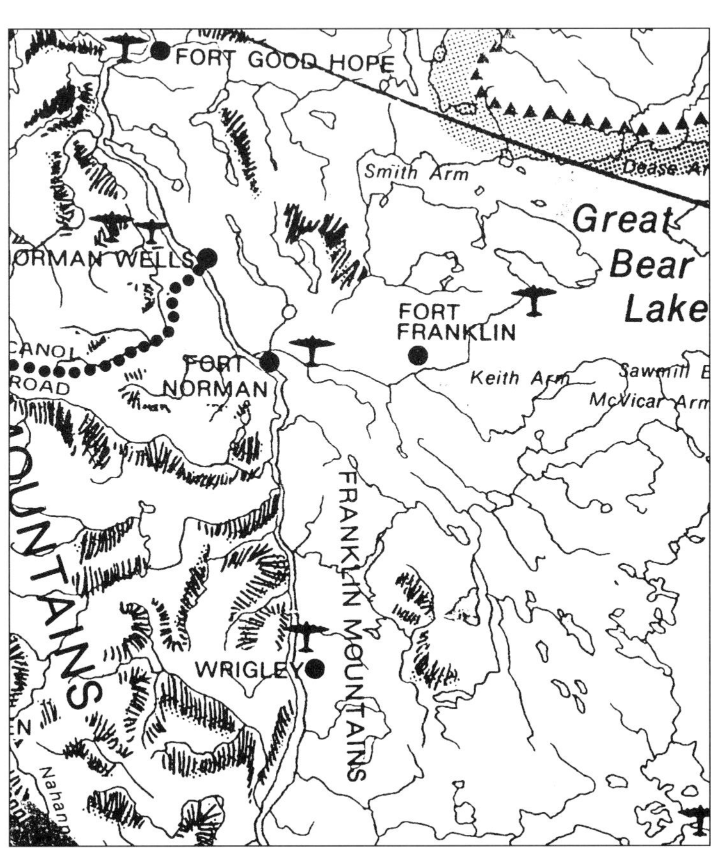

Figure 114. *Map showing location of Tulita (Fort Norman), "where two rivers meet." NWT Travel Map (1976)*

The following winter, game and fish nearby were scarce. Camsell and Allan Hardisty, the HBC clerk, traveled by dog team up the 128 km-long Great Bear River to Great Bear Lake. The Dene showed their appreciation for gifts of tobacco and tea by accepting in their small cabins the men, almost as family. It was a reverse kind of meeting, because usually those people came to Fort Norman to trade. Camsell and Hardisty fished for herring through holes in the ice using forked spears. After fattening their five dogs and harvesting a toboggan load of fish, they returned to Fort Norman.

In 1956 Dr. Bill Davies and I arrived from Fort Simpson. I do not remember much about the settlement. There were the HBC buildings, a home for the Department of Indian and Northern Affairs (DI&NA) supervisor and lumber for building him an office, an RCMP residence-and-office building, a nursing station with one nurse, a residence for two Northern Transportation Co. Ltd (NTCL) employees and several houses for other people, both Dene and outsiders.

Figure 115. *Old church in Fort Norman. Photo by Jack Grainge.*

Dr. Davies and I stayed at the nursing station. The Department of Public Works (DPW) had built the station and installed a septic tank and seepage pit. The subsoil was impermeable so that the nurse had to use a honey bucket and a wastewater pail. I felt sorry for the nurse whose nursing duties were difficult. She needed running water in her quarters and the nursing rooms.

I decided to help her. During the twilight night, I dug a trench the short distance from the seepage pit to the Mackenzie River bank. I made an outfall pipe using lumber intended for building the DI&NA building. I filled in the ditch, hiding my theft of lumber. I had not asked permission because I was afraid the RCMP might insist that I have official approval from some distant official, and I had to leave the following day. The nurse was delighted to heave the honey bucket and wastewater pail. Later the next day, the RCMP politely asked me to sign a statement listing the lumber I had used. I did but heard no more about my insignificant theft.

The septic tank effluent did not contaminate any water supply. In summer the Mackenzie River was turbid, so everybody in the settlement used clear water from Great Bear River. In the winter, the Mackenzie River was clear and the ice protected it from contamination, so the people used that water. At that time of year the effluent from the nursing-station septic tank froze on the river bank. The flooding river in May washed away the wastes.

The following year, the Imperial Oil Ltd. (IOL) medical doctor at Norman Wells, flew me in his two-seater plane to Fort Norman. The doctor could not accept payment for my transportation, but on my expense account I bought him a barrel of gasoline.

The water supply operator had a pump on a raft and a pipeline up the high river bank to a water truck. A remote electrical switch controlled the pump motor. He also hauled honey bucket wastes to a garbage dump east of the settlement. People discharged wash-water near their own well-spaced houses, where it seeped away.

I visited Fort Norman for my last time in the early 1960s. They had a school and teachers' houses. The water and wastes handling methods had not changed. I considered sanitary conditions to be reasonably satisfactory.

Two days before my visit, the sleigh dogs of the RCMP had broken out of their pen and savaged the HBC manager's six-year-old son. No plane was available to evacuate the boy. While traveling in a motor boat the 140 kilometers to Norman Wells, he asked to see the river once more. He died in his father's arms. They left the body with the doctor. The parents could not continue to live there. They flew out on the plane with me.

References

Camsell, C. 1954. Son of the North. New York, N.Y.: D. McKay Co.

FORT LIARD

Fort Liard is situated at the confluence of the Liard and Petitot Rivers. It is the oldest, continuously occupied, aboriginal site in the NWT. In 1805 the North West Company established a trading fort, Rivière aux Liards (river of farthings) there. A few years after the fort's establishment, sixteen or more Dene killed the traders. In 1820 the North West Co. reestablished the trading post, and a year later, with the union of the two trading rivals, it became part of the HBC.

Figure 116. *Location of Fort Liard. Department of Energy, Mines, and Resources Canada, 1965.*

In 1874 Charles Camsell was born in Fort Liard, son of Julien Camsell, the HBC factor. Julien had changed his surname from Onion to his mother's maiden name, Camsell. Father Grouard, later archbishop, the only other outsider, taught school for the Camsell children. At that time there were twelve to fifteen people there, six of them being Camsells. In 1884 Julien was made chief factor of the HBC Mackenzie River area, stationed in Fort Simpson. Charles, later Dr. Charles Camsell, was then six years old.

In small settlements such as Fort Liard, there was little danger of outbreaks of contagious diseases. Therefore we, in our office, did not usually visit them. We had a small staff and ninety or more percent of our work was in Alberta. However during the mid 1960s, a man in Fort Liard contracted typhoid. Dr. Gordon Butler, Director, Northern Health Services, asked me to go there to investigate the danger of a spread of that disease. I chartered a plane from Fort Simpson.

Upon arrival our plane circled south of the community. The half dozen, painted houses of the outsiders were framed by green grass on the southern slope of a high terrace in front, and green, scrub bush on the north slope behind. The glinting reflections of the sun on the river, as seen from the air, made a memorable picture.

A few days before I arrived, the RCMP officer had received word of the disease. He immediately advised everyone to leave the settlement. He told them to spread out with no two families together. He and his helper were the only people left in the settlement. I congratulated him for acting correctly. The RCMP personnel serving in the North receive training in medical emergency care.

Figure 117. *An RCMP Officer and his dogs in Fort Liard. Photo by Jack Grainge.*

The outsiders lived in a half dozen or so wood-frame houses situated on a gentle south-facing slope, slightly below the crest of a high gravel terrace facing the Liard River. The outsiders drew clear water from nine-meter-deep wells. During the summer, the Liard River water was silty. They had septic tanks and effluent seepage pits.

The Dene lived in tents and small houses on lower ground nearer the Petitot River. They used the clear water from that river. Everybody discarded wastes to the scrub bush area to the rear of the houses, near the tethered dogs.

The typhoid case was an isolated incident. The community was relatively clean. I was looking around the bush areas behind the outsiders' houses when a fierce dog jumped up at me and barked. I sprinted out of that line of tethered dogs, and that is another memory of Fort Liard I shall never forget.

FORT SIMPSON

Fort Simpson is situated on a delta island of the extremely turbid Liard River flowing into the clear Mackenzie River. In winter the Liard River water is clear. This turbid water in summer extends a half kilometer into the Mackenzie River. The flows of the two rivers remain not completely mixed for almost fifty kilometers downstream.

The land at the upper end of the community is twelve meters higher than the summer river level. The land slopes gently downward to the north, being about four meters lower opposite the dock. The two thirds of the land along side the Snye is much lower. It is subject to spring flooding and therefore was not occupied. The island, four kilometers by one kilometer, is separated from the mainland by a 150 meter-wide snye. At its upper end it is connected to the mainland by a causeway.

The soil consists of silt interspersed with layers of fine sand. The permafrost below the built-up parts of the community has thawed so the ground is stable.

Figure 118. *Party at Catholic mission, celebrating 100-year anniversary, 1958. Photo by Jack Grainge.*

For centuries the island had probably been a meeting place of Slavey Dene living along the shores of both rivers. In 1804 the North West Co. established a trading post there, which they named, "Fort of the Forks." Following the amalgamation of the two fur trading companies, the HBC renamed the post Fort Simpson, after George Simpson, the governor of the new company, with headquarters at Fort Simpson. Twelve years later he and the headquarters moved to Lachine, near Montreal. In 1841 he was knighted for his contributions to arctic discoveries.

In the early days the HBC-post employees cultivated a huge garden and raised livestock. The crews of the York boats, many of whom wintered there, needed the food. During their stay, the employees gathered firewood, fished, hunted and trapped. They made soap from wood ashes and in spring tapped birch trees to obtain syrup. The launching of the HBC, wood-burning, steam vessel, Wrigley, ended the York-boat era. This in turn spelled the end of the soap making, the syrup tapping and much of the gardening. Also the variety of foods sent to the trading posts improved.

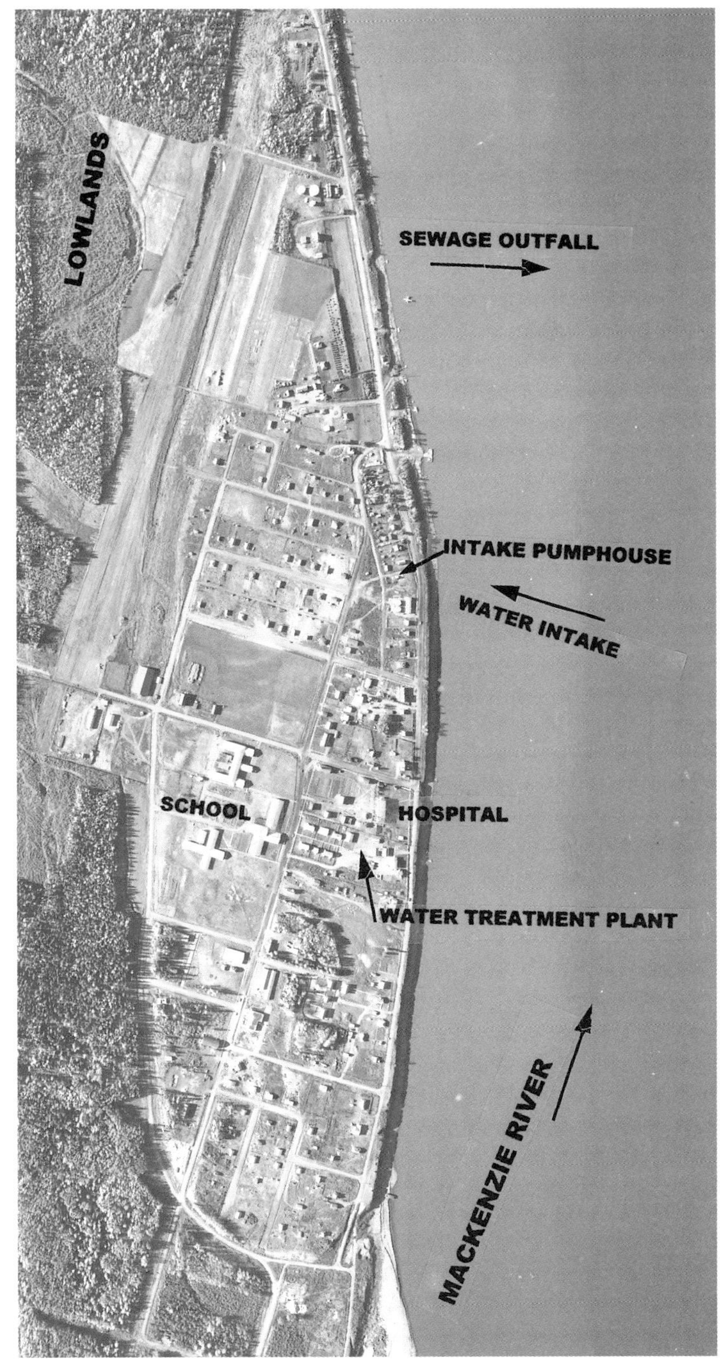

Figure 119. *Aerial photo of Fort Simpson. Department of Energy, Mines, and Resources, Canada, ca. 1961.*

In 1858 St. David Anglican Mission established there, followed in 1894 by Sacred Heart Catholic Mission. Anglican Bishop Bompas and Catholic Father Petitot made major contributions in the exploration of the surrounding region.

In 1882 Julien Camsell became chief factor of the HBC Mackenzie area and moved to Fort Simpson. He sent his children to school in Fort Garry, Manitoba. After graduation from a college in Winnipeg, Charles Camsell worked for the HBC at several forts. In 1897 he, together with his older brother Fred and a friend, headed for the gold rush up the Liard River. They spent two years in the Cassiar region and returned broke.

In 1910 Reverend Gerald Cord opened the first Indian agency. Flynn Harris took over the agency and opened a post office. In 1912 the Royal North-West Mounted Police established a detachment there. Until 1903 the name was North West Mounted Police. In 1920 the name was changed to Royal Canadian Mounted Police (RCMP). In 1916 the Catholic mission built St. Margaret's Hospital. In 1942, a half year after the USA entered the war, the American Army built a gravel airport on the mainland and a connecting gravel road, including the causeway from the Island to the mainland.

* * * *

In 1956 I flew to Fort Simpson in company with Dr. Bill Davies. We stayed in the local doctor's home

The buildings of the government employees, school and Catholic hospital ranged along high ground in an 800-meter-by-200-meter strip facing the Mackenzie River. The Dene preferred a wide section of lower land near the dock and the HBC post.

* * * *

The government employees and the hospital had water wells and sewage septic systems. The wells were about six meters deep, and all produced adequate quantities of clear water which the residents found to be satisfactory. Chemical tests we made showed high levels of certain nitrogen compounds in the water from the wells of the government employees, an indication of sewage pollution from the seepage pits of their septic systems.

The hospital sewage discharged into the Mackenzie River upstream of the dock where many people obtained water. Many others boated to obtain clear drinking water from the far side of the river.

The transport vessels at the dock pumped this contaminated water aboard. Drinking water passed through a Lynn filter, a manufactured, porous, diatomaceous stone, which removed the silt particles. When the filter became plugged, the cook turned an external handle which rotated a blade scraping away the silt along with some of the stone's surface. Then he flushed away the scrapings. It is likely that most of the bacteria would have become attached to silt particles and be flushed away. However we could not be sure that the drinking water was free of sewage bacteria. Wash water and dish water did not pass through the filter, and so might have been contaminated.

In 1951 Stan Copp had advised Father LeSage to build a septic tank for the hospital, with the discharge flowing to low land further inland. Due to a misunderstanding, his men constructed the septic tank in a low spot where Copp had intended the effluent from the septic tank to discharge. Father LeSage thought the system would not work so continued to discharge sewage into the river. I told Father LeSage that because wells were contaminated, I would be recommending public water and sewage systems with a sewage lagoon further down the interior of the island. That measure would solve both problems.

Figure 120. *Ted Trindle at Fort Simpson's water point. Photo by Jack Grainge (1972).*

In 1957 Norm Lawrence, the tall, husky, president of AESL, hired a small, dual-engine Cessna plane to fly the two of us to Fort Simpson. We hoped to jointly outline a plan for the water and sewer systems. We decided that clear water should be piped from the middle of the river.

We considered a sewage lagoon near the lower end of the island. The local administrator objected to that proposal because he said that the settlement would grow in that direction. I rejected the construction of a septic tank, because a flood would cause the septic tank contents to back up into the community. Two years later a flood covered that end of the island, so the community did not grow to the lower end of the island as predicted. A good site for a lagoon is near the snye. The annual spring floods would flush out the lagoon, at times when the water in the river is unuseable.

Because we could not agree on a suitable lagoon site, we decided that untreated sewage should be discharged into the river, well downstream of the boat dock. There would be a conveniently located, treated-water hydrant for the Dene, so that they would not need to obtain water from the river.

The untreated sewage would not pollute the drinking water. Therefore our proposed sewage disposal system met regulations under the NWT Public Health Act. With advice of consulting engineers, especially Ev Carefoot of AESL, I had prepared these regulations.

* * * *

In the fall of 1958 Poole Construction Co. (PCL) began construction of the water and sewer systems. Gordon Fuerst, a tall, alert, young, newly graduated engineer, represented AESL.

The water intake consisted of a 250-millimeter, steel pipe extending forty-five meters into the river. It was laid a half meter above the river bottom and held in place by a row of steel I-beams driven four meters into the river bed.

Water flowed into a nineteen-meter-high, five-meter-diameter, poured-in-place, concrete caisson, located close to the high river bank, and extending approximately two meters down into the river bed. Two alternately operating, submersible pumps in the caisson discharged water through a 200 mm, cement-asbestos pipe to the treatment plant, located a safe distance from the shore. Soon after the water system began operating, the motors below the impellers of the vertical turbine pumps became buried in silt. Uncooled by water, they burned out. They were replaced by water-cooled, vertical turbine pumps with motors above the impellers.

Figure 121. *Ice piling up along Fort Simpson shore. Photo by Jack Grainge (1972).*

The water treatment plant consisted of an oil-fired water heater, a liquid alum feeder, a mixing tank, a rapid sand filter, a gas chlorinator, and a treated water reservoir below the floor of the building. Backwash water from the filter discharged through a cement-asbestos pipe laid parallel to the intake pipe. It passed through the caisson and discharged into the river.

The water distribution system consisted of cement-asbestos pipe buried 2.5 meters deep. The mains were laid up and down the streets in a complete loop with water continuously circulating from the plant, around the loop and back to the plant.

Two, parallel, styrofoam-insulated, copper service pipes extend from the water main to each house. A continuously operating pump in each house circulated the water from one to the other. Because the sound of the pumps kept people awake at night, many householders shut off their pumps. They kept the water moving through their service pipes by allowing a small flow of water from their taps. This practice was costly because it wasted

Figure 122, 123, & 124. *The flood of 1963. Photos by Jack Grainge.*

treated water. Also some service pipes froze and had to be dug up and thawed.

The water distribution system, constructed in Fairbanks, Alaska approximately 1951, was the first recirculating, single-main, water distribution system. Likewise in that system, two parallel service pipes extend from the water main to each house. The two pipes are connected together in the house. Both pipes have horn-shaped ends, called fluistors, which project into the main. One horn points upstream and the other downstream. These fluistors keep the water flowing continuously through the service pipes. The operator circulates the water in the mains at a rate of a half meter per second. This system is quiet and less costly to operate than the one in Fort Simpson. I had explained the Fairbanks system to Lawrence, but he preferred to install a system with small pumps.

* * * *

In 1959, in company with Jim Anderegg and LaMar Hubbs, two engineers involved in research in my field of engineering in Alaska, accompanied me visiting settlements along the Mackenzie River. In Fort Simpson, we stayed a night in the hotel constructed and operated by Mr. Kidd. On the main floor was a restaurant and an office. Upstairs there were six bedrooms around the perimeter with the stairway and a honey bucket in the middle. I forget details of the washing facilities, but I remember that the honey bucket was not flushed during our stay. I understand that later Kidd improved the hotel.

Figure 125. *Dogs remained in Fort Simpson during the 1963 flood. Photo by Jack Grainge.*

In May 1963, there was a disastrous flood on the Mackenzie River at Fort Simpson. Due to a late, rapid thaw in the headwaters of the Liard River, the flood was unusually heavy. A huge pile of ice chunks smashed into the solid ice covering the Mackenzie River. The ice chunks formed a dam across the mouth of the river and extended thirty meters along the shore of the Mackenzie River. Chunks of ice were pushed up on the shore, higher than the tops of the power poles, approximately twenty meters higher than the usual water level.

I arrived at Fort Simpson after the people had been evacuated by airplane to Yellowknife. Only twelve people remained on the island. They had been responsible for getting the people to high ground at the school. They cooked and served dinners to the people and when planes were available, drove them in relays to the airport.

After the people had gone, the twelve men continued to work, cleaning up various buildings. I spread a chlorine compound on the ground around the manholes which had backed up. I helped the operator get the water plant producing somewhat clear water.

The huge chunks of ice from the Liard River ripped out the water intake pipes and the steel piles that had held them in place. The shortened water pipes continued to operate, but the water pumped was muddy. The backwash water from the filter discharged near the water intake, and probably some of it was pumped through it back to the plant.

The subsequent construction of a buried, water-intake pipeline to clear water on the opposite half of the river is a significant achievement. Burying it among boulders too heavy to be moved by floods is costly, but saves a fortune in water treatment.

References:

Camsell, C. 1966. *Son of the North.* New York, N.Y.: D. McKay Co.
Feurst, G. 1996. Personal communication.
Lawrence, N. 1996. Personal communication.

YELLOWKNIFE

Detah was the first permanent settlement of Dogribs in Yellowknife Bay. It is situated on a rocky point, on the east shore at the mouth of the Bay, six kilometers downstream of Yellowknife. The Dene chose the site because it was not near mosquito-breeding swamps and lichens.

In 1929, bush-pilot Stan McMillan flew a party of prospectors to Yellowknife Bay. He reported five log cabins at Detah and none on the present site of Yellowknife. When I first went to Detah in the late 1950s, there were twenty or more Dene there and several on nearby Joliffe and Latham Islands. All were living in small log houses.

Figure 126. *Aerial photo of Yellowknife. Department of Energy, Mines and Resources, Canada, 1979.*

Figure 127. *Con Mine and Pud Lake, Yellowknife. Photo by Jack Grainge (1976).*

Latham Island was named after Gordon Latham. He was a pilot for Canadian Airways, first at their school for RCAF navigators in Edmonton, and later at Yellowknife. Also he had been a partner in the first hotel and restaurant, originally in a crowded tent, on that island. Due to a long bout of sickness he lost his share. In 1936 Latham helped Tom Payne stake a claim on a narrow gap between the properties of Con and Negus mines. The night before the claim held by Con Mine lapsed, they rowed to the mainland. At midnight they started staking and finished before another party arrived. Con Mine's stakers came later in the morning. Both Payne and Mickey Ryan of Fort Smith, who had been grubstaking Payne for three years, became rich overnight.

In 1964, Chief Sangris and his people in Detah chose Joe Toby, one of their respected young men, to train as their Department of National Health and Welfare (DNH&W) community health worker.

Figure 128. *Joe Toby and daughter in their home. Photo by Jack Grainge (1976).*

Figure 129. *Con Mine, Pud Lake and tailings pond; Kam Lake in Background, May 1976. Photo by Jack Grainge.*

Figure 130. *Giant Mine, Yellowknife. Photo by Jack Grainge (1959)*

He worked with the public health nurses and our engineers. Toby organized a wastes disposal area on the east-facing slope behind Detah. He also explained to his people the public health nurses' advice. The public health nurses were in turn Barbara Stanton, Joe Atkinson, Sister Mopati, Lillian Piper, Ann Wheeler and Mike Hewitt. A few years later Toby became an interpreter-announcer for the CBC. Eventually Hewitt became Deputy Minister of Justice.

In 1950, a Canada-wide polio epidemic struck Yellowknife. Barbara Bromley, a regular volunteer in clinics for babies and immunization, helped transform the medical clinic into a major immunization center. In 1967 her husband Peter and dentist Doctor Ian Calder drowned when they were boating down the Back River. Her son Bob swam to shore completely exhausted. Until he was rescued nine days later, he lived on three energy bars and roots of water plants.

In the 1950s, after the Yellowknife sewer was operating, Stan Copp and I started a program of sampling the water near Detah and the southeast end of Latham Island. For several years the water was safe for drinking. In the late 1960s, sampling showed that the water was contaminated. Undoubtedly that was because Niven Lake, the city's sewage lagoon, had become too small to serve the growing population of Yellowknife. There was a delay in switching to using Kam Lake. In 1976 the federal government constructed ten kilometers of road so that treated water could be trucked to the people in Detah. At the time the road was a series of potholes, but the water trucks made it.

*   *   *   *

During 1935-36, Burwash Yellowknife Gold Mines Ltd. mined out a pocket of rich gold ore on the east shore of Yellowknife Bay and then closed.

In 1936, prospectors for Con Mine, Negus Mine and Giant Yellowknife Mine staked gold claims. They immediately began large scale construction of their mines and accommodation for their workers. The "Old Town" of Yellowknife grew on both sides of the narrow strait between Latham Island and the mainland. Within a year it became a haphazard confusion of shacks and tents. At the rise in the road near Weaver and Devore Store, the only two cars in Yellowknife collided head on.

By 1938 "Old Town" had grown along a couple of streets to become a jumble of log cabins, and wood-frame shacks, houses, stores, offices, a bank and two restaurants. They spread out from the rock peninsula to the flatland to the south, and to the nearby shores of Jolliffe and Latham Islands. The population was about five hundred, up from one hundred the previous year. Con and Giant mines each had about two hundred people living on their sites and Negus Mine had less than one hundred.

In 1938-39 Con, Negus and Giant mines began producing gold. Due to a wartime shortage of manpower, all of them ceased milling in 1942 or soon thereafter. However some mining continued. After the war all of them resumed full operations.

*   *   *   *

Yellowknife was a hub for planes from Edmonton serving mines and communities in the Western and Central Arctic. Pilots landed planes on Yellowknife Bay and docked at the Canadian Airways base on the mainland in Old Town. In 1943, Fred Miller came to Yellowknife as base manager for Canadian Airways. Ed Bolger, manager of Eldorado Mine warned him: "You hire a blankety-blank pilot who can get supplies into our mine or I'll hire a pilot with an airplane and to Hell with you." Canadian Airways transferred Ernie Boffa to Yellowknife from Regina where he had been flying planes for airforce navigators who were in training. The same year Canadian Pacific Airlines bought Canadian Airways.

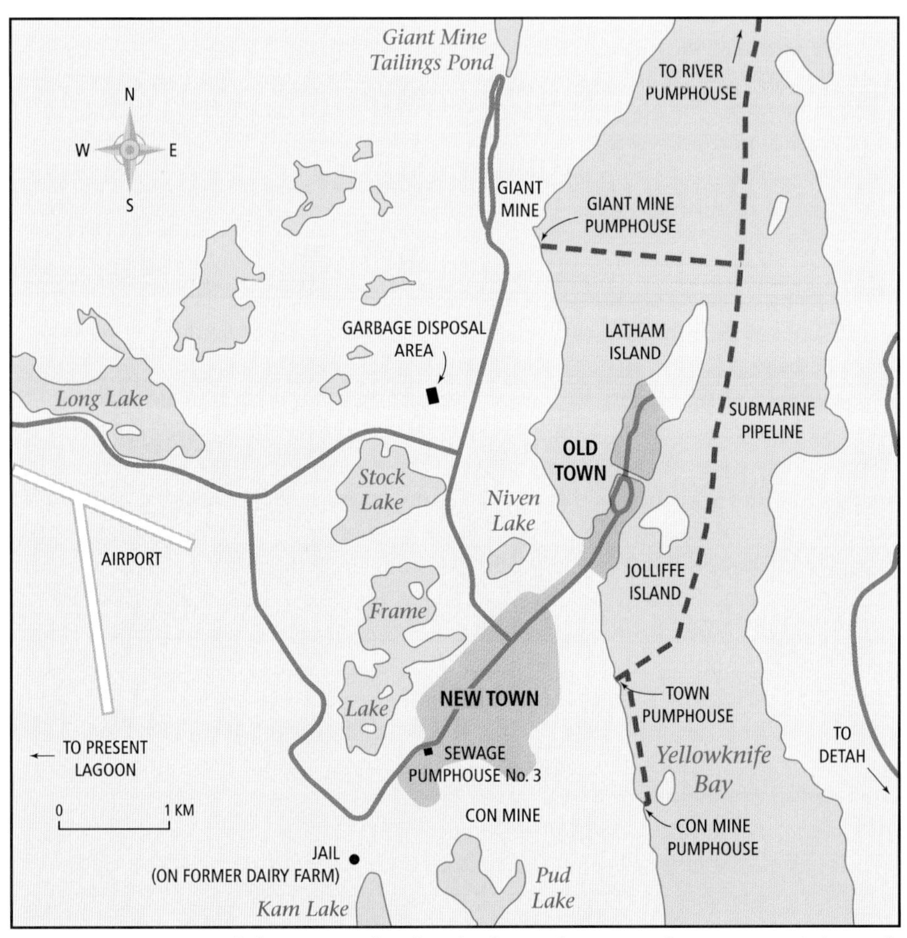

Figure 131. *Map of the town of Yellowknife and surrounding areas. Adapted from EPEC Consulting Western Ltd. (1981) by Johnson Cartographics, Inc. Edmonton.*

Boffa soon had occasion to demonstrate his inventiveness in getting through. Father Gathy received a radio wire about a medical emergency. A priest at Repulse Bay on Melville Peninsula had frozen his arms and needed to get to a hospital. Presumably planes from Fort Churchill, which is closer, were not available. Boffa filled his single engine Norseman's gasoline tanks, and strapped four, two hundred-liter drums full of gasoline inside the plane. He taped the nozzle of the filler hose so that it fed into a wing tank. Father Gathy sat in the back with a toggle pump in one of the barrels. On signal from Boffa, he fed gasoline to the wing tank. Little of the land between Yellowknife and Repulse Bay had been mapped, and Boffa was unfamiliar with the area. Referring to strip maps, he flew an indirect route.

On his return from Coppermine in 1948 with Duke Decourtney, owner/operator of a local newspaper a passenger, Boffa's plane began to descend on its own. Boffa landed the plane on a lake, cleared a fuel line and continued on his way to Yellowknife.

Figure 132. *Aerial view of Yellowknife, showing Old Town and Latham Island. Photo by Jack Grainge.*

Athletic Fred Miller and his pretty wife, Gerty, were mixers. On winter evenings they played badminton at Con Mine's recreation hall. During the depression, years earlier, he hopped on a freight train, heading east from Winnipeg. He ended up on a road construction crew in Ontario, where he spent his evenings boxing in a ring. His skill became known and that winter he became a lumberjack. In the evenings he was transported from one camp to another to win the championship for his camp. If he had not been a good boxer, he would not have had a job. Such were the realities of living during the depression.

Years later I met Miller at the Royal Glenora Club in Edmonton. He played badminton with me, but I never attained his class. He was a key man whenever tournaments required an organizer.

Miller and Bing Rivett of the RCMP, ran the boys' baseball league. Miller, Rivett and other friends sang around bank manager, Charlie Desson's piano, the only one in the settlement. One Christmas, Miller, Rivett and Fred Breau sang in two church choirs. They sang in the Anglican Church for Reverend Randall, then at midnight mass in the Catholic Church for Father Gathy. On the cold ride from one church to the other, they warmed up with over-proof rum. At the Catholic Church they were grouped near the wood-burning heater at the back of the church. Fred Breau, the only Catholic, prompted them for the responses. The over-proof and the heat were too much for Breau. He fell asleep. The choir woke up when Father Gathy chanted, "Freddy, you may sing now. Freddy, you may sing now."

In 1945 or 1946, Miller, Ivar Johnson, a local building contractor, Blacky Blackburn, the DOT superintendent, and Ted Cinnamon, contract wholesaler for IOL, built an airstrip at the site of the present airport. Cinnamon had a small Caterpillar tractor with a narrow front blade. With several other men wielding shovels and axes, they cleared away bushes, scraped away humps and filled in dips. That night they toasted their good work with Miller's over-proof. Perhaps that was why they did not notice that the runway was swaybacked.

Canadian Pacific Airlines flew in a twin engine, low wing, Barclay-Grow airplane. It touched down a short distance before the swayback dip. At the beginning of the rise, the tips of the propeller furrowed the ground. Ivor Johnson hammered the propellers straight enough for the plane to fly back to Edmonton. Later a DOT crew rebuilt the airstrip.

Canadian Pacific Airlines built a small office and waiting room at the airport. They began daily flights of a Barclay-Grow aircraft from Edmonton, landing at Fort McMurray and Fort Smith en route. The smartly dressed crew stayed at Vic Ingraham's Old Stope Hotel.

Yellowknife also became a hub for flights to mines and settlements in much of the Western and Central Arctic. In 1945, Gerry and Norm Byrne prospected Consolidated Discovery Mine on Giaque Lake, a hundred kilometers north of Yellowknife, and immediately developed it. The mining equipment was barged to Yellowknife. From there, John Dennison hauled it to Discovery over an ice road, which he, with help from Stu Demelt and other tough northerners, had constructed.

During the summer of 1953, I had an interesting flight from Yellowknife to Discovery Mine. Steve Homulos, the area mining inspector, had advised Stan Copp, my boss, that he suspected that the mine's water supply was contaminated by the mine's septic tank effluent. I planned to investigate the problem, and after doing so, I advised a diversion of the effluent.

I was waiting at Max Ward's dock in Yellowknife while Ward loaded his single-engine, Otter aircraft for the flight to Discovery Mine. Up walked the area RCMP Inspector. He told Ward that RCMP were checking everyone along all roads on which Tony Gregson, a thief, might be escaping. He said it was a net from which Gregson could not escape.

On our hundred kilometer flight to Discovery, Ward told me that Gregson had carefully planned the theft of two gold bricks from Ward's cargo. He timed his getting fired at the Mine with the weekly Friday flight on which two gold bricks were being shipped to Yellowknife. Gregson had previously hammered two lead bricks to the same size and shape as the Mine's gold bricks. He had also made a bag of the same material and size as the bags in which the gold bricks were shipped.

On his way back to Yellowknife, Ward stopped at the Mine's wood-cutting camp. While he was outside unloading some groceries, Tony substituted his bag of lead bricks, complete with a faked shipping tag, for the Mine's bag of gold bricks.

At the dock in Yellowknife, Ward helped Gregson with his bag. Ward exclaimed, "Wow! What a heavy bag! What have you got in it? Gold bricks?" Gregson snapped, "Be careful with that bag. It contains my radio."

Tony then hired Ward's Beaver to fly him to Hay River so that he could catch the next bus to Peace River, Alberta.

Apparently Gregson knew that the bag containing the substitute bricks would stay in Ward's office until men from Frenchie's Transport would collect it on the following Monday. When they did pick up the bricks, their shipper noticed an error on the shipping tag, and called the RCMP.

On our return to Yellowknife later that day, I learned that Gregson completed his escape with $70, 000 worth of gold, a lot of money in 1953.

Ward studied the flying services market carefully. He reported in his book, The Max Ward Story, that many pilots in Yellowknife predicted that he would be unable to keep up the payments for his hundred thousand dollar Otter. However he won contracts for hauling large equipment and heavy loads that the competing companies, with their smaller planes, could not handle. He eventually had a fleet in Yellowknife of five Bristols as well as single-engine Otters and Beavers.

Ward's Bristols could carry large equipment, such as a D4 bulldozer, for long distances. Thus he made it possible for geologists Gerry and John Byrne to develop distant mines such as Taurcanis, three hundred kilometers northeast of Yellowknife and Rayrock, seventy kilometers north of Rae.

Other airlines in Yellowknife flying smaller planes enjoyed a brisk business of flying people to mines and northern settlements. The wily Tommy Fox's Associated Airways bought Canadian Airways base and planes. In 1953, when I flew to Coppermine with Ernie Boffa as pilot, Merv Hardy was the manager of Associated Airways base. The next year Hardy won the election to become the first Member of Parliament to represent the NWT. At other times I flew with other pilots who ran single-engine, one-plane businesses, Jim McAvoy, Henry (Hank) Koenan, Gordon (Smoky) Hornby, as well as Chuck McAvoy, who disappeared on a trip and was never found.

Ptarmigan Airways had a base on the south side of the peninsula. Bill Jewitt, flying the Con Mine prospecting plane, used the Ptarmigan dock. So did Koenan.

* * * *

During Yellowknife's early days, people bucketed water from Yellowknife Bay from whatever shore was closest. After a few cases of typhoid had occurred, Con Mine's Dr. O.L. (Ollie) Stanton, who arrived in 1937, declared the water in Back Bay to be unfit for drinking. He said water should be drawn from Yellowknife Bay, well offshore on the southeast side of the peninsula. In 1941 Tjart (Tom) Doornbos, a stubble-bearded Dutchman, arrived. With a drip hanging from the end of his nose and a bucket hanging from each end of a shoulder yoke, he became the first public water supply system for "Old Town." Of course many people carried water for themselves.

Gertie Miller, wife of Fred, the manager of CPA became the self-appointed, unofficial public health inspector. From her house high on the peninsular hill, she would watch to see whether Doornbos blew his left or right nostril. She would then run ahead to warn her friends which bucket was unacceptable. Dr. Ollie Stanton advised people to boil their drinking water and to get TABT shots.

Figure 133. *Part of old town showing solid rock which could not absorb wash water. Photo by Jack Grainge.*

Doornbos charged 25¢ per bucket and worked long hours. Perhaps his was one of only a few profitable, water-supply systems in Canada. Obviously he did not spend his profits on clothes. Rumors of his riches from investments abound. He bought several building lots in both Old Town and New Town. He also staked a few claims for investment dealers in Toronto. He told me that the dealers did not care whether the claims were good or not. They just had to be near good claims.

I heard a popular rumor about Doornbos. An employee of Con Mine chased him away from the Company's junk pile. Later the mine manager received a letter from head office telling him not to bother one of their major shareholders. Doornbos told me he traveled extensively in Europe, probably on the cheap. His fame spread and during the 1980s; his photo made the cover page of *Maclean's*.

In 1939, Johnny Baker, the federal mine recorder, with several other government administrative responsibilities, hired men to construct a public, summer water distribution system. The peninsula, a 45 000-liter, wood-stave, cylindrical tank on the thirty-meter-high peak of the peninsula's hill and galvanized iron distribution pipes of sizes up to fifty millimeters. The pipes ran on the surface of the ground to houses and hydrants on the peninsula. There was no charge for using this water.

In 1946, the City began a winter, "trucked-water" service. A few people had indoor water reservoirs, but most used abandoned 45-gallon oil barrels.

* * * *

During the 1930s, people threw wash water all over the place, wherever neighbors least objected. Soon after Dr. Stanton arrived in 1937, he stated that the disgraceful, unsanitary conditions must be cleaned up. He disallowed pit outhouses on the rocky parts of the settlement because there was too little soil cover. Most people used honey buckets. People, with houses situated on soil, installed outhouses with locks on the doors or septic tank systems. Dr. Stanton ineffectively ordered people to boat their garbage to land across the Bay.

Many people were clamoring for public outhouses, and finally, in 1938, the mining recorder hired a carpenter to build two, with buckets. Within a few days they were too filthy to use. No one would clean them. Finally someone burned them.

Figure 134. *Wildcat Café, where wash water discharged to open ground. Photo by Jack Grainge.*

In 1938, Vic Ingraham built the first, of his three, consecutive, large, two-storey, wood-frame hotels. Ingraham was a hard-driving winner. After enduring severe burns when his schooner on Great Bear Lake caught fire, he and his helpers drifted to shore on a rubber life raft. He froze the ends of his fingers and his legs, which had to be amputated. Friends sometimes called him 'Peg Leg.' His first hotel was situated high on the rock on the peninsula. A steady stream of sewage from that building flowed through the middle of the settlement and into Back Bay. There were also four restaurants discharging wash water outside their premises, two on each side of the Peninsula. During 1946 and 1947, the Trustee Board discussed that problem. They passed a bylaw requiring cafes to have their wash water hauled away, but none of them did so.

In 1947 Dr. Stanton identified two cases of typhoid. He contacted Allyn Richardson, one of my predecessors. Richardson recommended that above-ground sewers be laid to discharge the wastes from restaurants into the Bay. In summer this system helped, but did not solve many other problems.

This unsanitary situation continued until 1949 when the New Town water and sewer systems began operating and the restaurants in Old Town closed. In 1949 Vic's Old Stope Hotel burned down.

Young dogs, running loose, scrounged on discarded wastes throughout the settlement. Many people objected to the dog manure throughout the settlement. However dog owners argued that if pups were not allowed to run about, they would not develop into hardy sled dogs. Grown dogs were staked wherever there was earth on which they could lie comfortably.

With a two-horse team and either a sled or wagon, depending on the season, Frank Buckley hauled barrels of toilet wastes and garbage to a site at the west end of Niven Lake. In winter he turned the barrels upside down on the frozen ground and waited for the summer thaw. Buckley sold out to Tom Reed, who carried on the profitable business.

In later years the Town operated a garbage disposal in an area west of the road to Giant Mine. During the summers the workmen bulldozed garbage together and set it on fire. In later years they buried the remains. It was a reasonably clean operation so that flies, gulls and ravens as well as surface water run-off were not public health problems. At the time it was consistent with generally accepted standards. Discarded cars, machinery and other metal scraps were heaped separately. They were not buried.

* * * *

In 1944, lawyer Fred Fraser, Department of Mines and Resources, came to Yellowknife. He was stipendiary magistrate, mine recorder and administrator of everything else that needed administrating. He recommended the construction of a new town on the elevated plain west of the old town. Unbelievably there was some opposition. Jock McMeekan, publisher of the *Yellowknife Blade,* referred to the proposed new town as "Fraser's Folly" and "Blunderville."

Straight-forward Fraser, an experienced administrator, knew how to handle opposition. He was fair, but strongly advocated the construction of a new town. He recommended hiring Professor Norman M. Hall, Mechanical Engineer,

Figure 135. *Relaying above-ground water mains. Photo by Jack Grainge (1962).*

University of Manitoba, to design the water system. Professor Hall had previously designed the water and sewerage systems for Flin Flon, a lead-zinc-copper mining town in northern Manitoba.

The water distribution and sewer pipes at Flin Flon were contained in ninety by ninety centimeter, unpainted, above-surface wood boxing, called utilidors. They were filled with sawdust insulation. The tops of the utilidors were waterproofed with a coat of asphalt, and people used the utilidors as the sidewalk system. To prevent the water and sewage from freezing, it was heated in the water pumphouse. Con, Negus and Giant mines in Yellowknife and other mines used utilidors, approximately fifty by fifty centimeters.

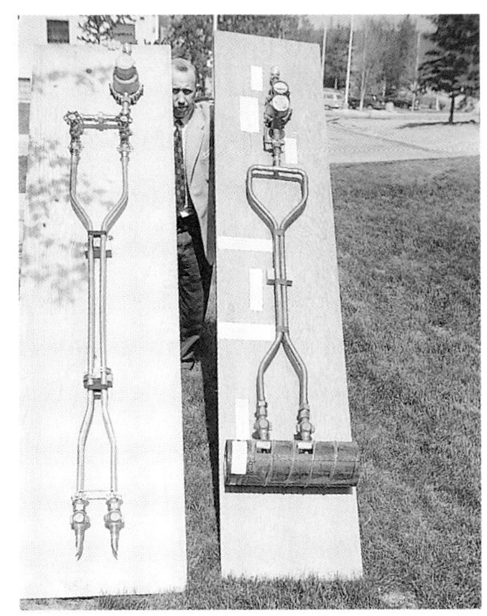

Figure 136. *Both ends of service connections, showing how the Fairbanks system worked. Photo taken in Fairbanks, by Jack Grainge (1962).*

Hall decided that such a system would be too costly for a town of 5000, the ultimate population for which he was required to plan. He conceived a clever idea, a dual-main, water distribution system. In the water plant, heated water would be pumped to a system of high-pressure mains and service pipes. Inside each house water would flow from the high pressure service pipe, through a tiny orifice in a pipe connection, to the low-pressure service pipe. Water would flow back to the plant through a system of low pressure mains running parallel to the high-pressure mains. At dead ends and fire hydrants, pipe connections with tiny orifices ensured a constant flow of water from the high-pressure to the low-pressure mains.

As mentioned in the chapter on Fort Simpson, engineers for Fairbanks, Alaska, subsequently designed a single main, constant flowing, water distribution system. Obviously the single-main system was quieter, better and much less costly to both construct and to operate.

Figure 137. *Relaying buried dual water mains. Photo by Jack Grainge (1962).*

During the 1960s, Reid Crowther and Partners, consulting engineers, also began designing circuits of single-main, circulating water systems for new subdivisions in Yellowknife. They planned the system with a continuously operating, small pump in each house drawing water from the main by way of one service pipe and discharging it back to the main through another service pipe. It was the same system as the one installed at Fort Simpson.

I explained the successful Fairbanks system to senior consulting engineers for both Yellowknife and Fort Simpson. They both thought the system with pumps would be better. The Fairbanks system was both better and less costly.

* * * *

Northern Construction Co. and J. W. Stewart, of Vancouver, constructed the water and sewerage systems in Yellowknife for cost plus 10%. John MacNeil became the project superintendent with Slim Locum his foreman.

Engineers of the Edmonton office of the federal Department of Public Works (DPW) made the surveys and supervised the construction. Ed Kellet became the chief with John Piard in charge of surveying. Work began in 1945 and was completed in 1949.

The first step in the laying of the water mains and sewers was to bulldoze away the bush. Then, to thaw the permafrost, they bulldozed away the moss and grass along the routes of the pipelines. Work could not proceed until the permafrost had thawed. Then the water-saturated soil prevented the laying of the pipes. Therefore the workmen laid a drainage pipe system below the intended sewers. It consisted of 150 mm asphalt-coated, flexible, spirally wound, galvanized iron, bottom-perforated, drainage pipes. They laid the drainage pipe so that the sewers could be laid directly on top of them. This drainage system discharged into Yellowknife Bay downstream from the sewage treatment plant. I recall that the system continued to drain for several years after construction.

The water mains in Yellowknife were buried deep. The sewers were located 1.8 m horizontally away from the mains, and slightly deeper. The mains were Anthes Imperial Co. cast iron pipes. They had flanged ends by which they were bolted together. This type of pipe is simple to assemble, but the bolts and nuts are prone to corrosion. At Yellowknife they lasted surprisingly well, approximately thirty-five years.

Before backfilling the water and sewer trenches, the workmen scattered forty centimeters or more of moss into the trenches for insulation. During the winter of 1951-52, Stan Copp, C.B. Crawford, geophysicist, National Research Council and I made a small research study of permafrost. We found that because the moss around the pipes had become water soaked and then frozen, it had negligible insulation value.

The water-treatment plant was situated in the School Draw, near Yellowknife Bay. Two service pumps in a deep, waterproof, concrete dry well discharged water to the supply line and also to a large water reservoir in the plant. A Wallace and Tiernan gas chlorinator injected chlorine into the water. A fire pump, also in the dry well, pumped water to the water supply line. A standby W & T gas chlorinator was connected, ready to inject additional chlorine whenever the fire pump began pumping.

About five years later, upon our recommendation, the Superintendent of Public Works installed a fluoridator. Initially we required a fluorine concentration of 1.0 mg/L. During the 1960s, Dr. Gordon Butler, Director of Indian and Northern Health Services, DNH&W asked me to study fluoridation as related to people in the North. I investigated the best concentration of fluoride in the water to provide protection of teeth from cavities without causing dark mottling.

At the time the recommended level by the U. S. Public Health Service varied from 0.7 mg/L for the southern states to 1.2 mg/L for the northern states. The difference was based on the estimated average amounts of water people would be drinking. Levels were to be adjusted based on dental observations. The Alberta standard continues to be 1.0 mg/L.

After discussions with Dr. Butler, we recommended a concentration of 1.5 mg/L. The water supply operators at Yellowknife and Inuvik followed those instructions and the dentists have never suggested that the figure was too high.

The outgoing water from the treatment plant was heated by a steam heat exchanger to a temperature of 5°C. In both summer and winter it returned about a half degree cooler. To save both heat and the pressure in the return water, it together with make-up water from the reservoir, was pumped by a recirculating pump into the high-pressure main.

After each use of a fire hydrant, it was sucked as dry as possible and a small amount of antifreeze added. Years later, a public works superintendent claimed to be afraid that one of his men might mistakenly add enough antifreeze to poison someone using the water. Therefore he regularly bought ethyl alcohol at the liquor store. He claimed he used it instead of antifreeze. The authorities trusted that he kept the alcohol under lock and key. He did not offer me a sample.

It is difficult to imagine a combination of circumstances that would result in the antifreeze working its way through a closed valve, against the pressure of the water in the mains. However I suppose the superintendent told the liquor store manager a good story. Evidently the superintendent's successors were less convincing to the liquor vender, because the former practice of adding antifreeze was re-instituted.

* * * *

The first sewers were asphalt-coated, flexible, spirally wound, galvanized iron pipes. They were laid directly on the drainage pipes 1.8 m from, and 60 cm lower than, the water mains. The joints between the pipe ribs were thin. Consequently they corroded, eventually leaving only the ribs. The pipes then looked like a row of fish skeletons. During the 1960s, the superintendent began replacing them with asbestos-cement pipes.

A standard primary sewage treatment system was part of the original construction. It was located along the lake shore a hundred meters or so downstream of the water treatment plant. It consisted of a screen, settling-and-skimming tanks to remove sludge (solid particles) and a heated, sludge digester. The liquid effluent was pumped from there, through a 150 mm, wood-stave pipe to the second pump station across Franklin Road, and from there to Niven Lake.

At first Niven Lake was approximately five hectares in area. Two years later the Superintendent of Public Works built a dam across the outlet to Yellowknife Bay, which increased the area of the Lake to eight hectares.

In about 1956, when I was examining the sewage treatment plant, Roy Cinnamon, the operator, pointed out the high cost of labor and heat. Since treatment of sewage in lagoons had become common throughout western Canada and USA, I suggested that sewage be pumped without treatment directly to Niven Lake. At the time the population of New Town was approximately twelve hundred. The area of Niven Lake is eight hectares.

Effluent from Niven Lake overflowed into Back Bay and people living on Latham Island sometimes used untreated water from the Bay. By the time it reached the water-supply intakes of Con Mine and the Town, the effluent from Niven Lake had become mixed with, and diluted in, the water of Yellowknife Bay. However there was a direct flow through the strait between the mainland and Latham Island. Therefore reliable operation of the chlorinators of Con Mine and the town was essential. For a price of five dollars per house per month, a town employee delivered treated water to houses on Latham Island. However some of those people bucketed untreated water from the Lake.

Considering the Town's small population at that time, I considered that it would be impractical to pump sewage to any lake other than Niven. However our studies in approximately 1967 showed that Niven Lake was too small.

* * * *

During the first few years after construction, DI&NA operated the water and sewerage systems. Ed Lundman, who had been the assistant DPW engineer, became the first Public Works Superintendent. He kept the water and sewerage systems operating well.

Figure 138. *Houses constructed on unstable ground that was underlain by permafrost. Photo by Jack Grainge (1962).*

After serving two years, Lundman resigned. His replacement became fed up with frequently having to dig up and repair breaks in the trunk mains running from the water plant to the town. The breaks were probably due to the melting of permafrost in the unstable soil. He relaid that section of both water pipelines on the ground surface, and heaped a meter depth of moss over them. The dry moss was a satisfactory insulation. Years later a foreman re-buried the pipeline with no further problems occurring. Apparently the permafrost had melted. About fifteen years later, many houses built near there did shift topsy turvy.

In 1953, Yellowknife became a municipal district with an elected mayor and council. The Council took over the operation and maintenance of the public works.

The next public works superintendent was an imaginative engineer from Ireland. He constructed living quarters for his family on the second floor of the city hall. I thought that he was competent. However, like all of us, he made a few mistakes. He considered that lowering the level of Frame Lake would lower the too-high, subsurface water table in Yellowknife. He pumped water from that lake for a month without affecting the lake level. He concluded that the pump was too small, and that lowering the lake was not worthwhile.

He accepted a position as assistant engineer somewhere in British Columbia. He and his family visited with my wife and me on their way through Edmonton en route to his new job. We enjoyed their company.

The next public works superintendent replaced the water pumps and motors located at the bottom of the concrete well with well-pumps. He set the motors higher than the treatment plant floor. He assured the Mayor and Councillors that they did not need to retain a consulting engineer because he himself was capable of performing engineering work. Later he made costly mistakes and resigned.

In 1957 the Town Council hired Jo Leiro as public works superintendent. I introduced him to a friend, who seemed puzzled by his name. The tall, slim, blond, Nordic Leiro commented, "I'm Italian."

In Norway, Leiro had graduated as a carpenter from a trade school and subsequently as a municipal works technologist from an engineering technology college. Although he did not have an engineering degree, he was a well-educated, experienced, municipal technologist, a great asset for Yellowknife. We often discussed the Scandinavian engineering approaches to municipal engineering problems as related to northern climates. For example, Scandinavian engineers plan towns so that there are a minimum number of sewage pump stations, which are costly to both construct and maintain.

In 1962, Leiro showed his skill. He relaid the outfall sewer eliminating one of two pump stations. All sewage then flowed by gravity sewers into a tank at the pump station halfway down the hill to Old Town. Thus the land on the northwest side of Franklin Road could be served by gravity sewers leading to that pump station. I thought of buying some lots in that area, but did not do so because it would have meant a conflict of interest. I would have made a bundle.

Leiro asked the Town Council to hire Haddin, Davis and Brown Ltd., later to become Reid Crowther and Partners. They planned new streets with water mains and sewers. Rod Tweddle became their first resident engineer. He and his family rented a house in Old Town, next door to his congenial landlord, John Anderson-Thomson, a consulting engineer and surveyor. He was one of Yellowknife's pioneers. When introducing himself, he sometimes punned, "There's no 'p' in 'Thomson'."

Leiro was an efficient superintendent. Unfortunately his wife, Kirsten, felt isolated in Yellowknife. Her mother and father helped by coming from Norway and staying two years with them. Kirsten worried too much when workmen, whom Jo had fired, would call by telephone and tell them to 'go back to Norway'. She, her parents and Britte, their pretty, seven-year-old daughter, did go back. Leiro stayed on for another few months but became lonesome. In about 1963, he returned to his home town of Dale, Norway. Yellowknife lost a superior superintendent. With Councillor Ted Cinnamon's cooperation, he had made many improvements in the town.

Leiro became the engineer for a district consisting of Dale and two nearby towns. He and I and our families have remained good friends. He and his family have stayed with us in Edmonton. My son, my wife and I have twice been welcomed in his home in Dale. In the mid 1980s, Leiro retired and soon thereafter Kirsten died. In 1994 he and his second wife, a Canadian from Montreal, visited me in Edmonton. They live sometimes in Montreal and at other times in Norway.

Fritz Thiel, a humorous, cheerful fellow, formerly in the German army engineering corps, replaced Leiro as public works superintendent. He was a capable engineer and Mayor Fred Henne offered him strong support. In 1972, Jack Kraft, the assistant engineer for the town of Esterhazy, Saskatchewan, became Thiel's assistant. Stu Demelt was his capable foreman.

* * * *

Since 1949 arsenic from Giant Mine was slightly contaminating the water in Yellowknife Bay, the source of water for the town and the mines. In 1966 Yellowknife was on the verge of a population boom, so we recommended piping arsenic-free water from the Yellowknife River, upstream of the Bay, to the water supplies of the Town and the mines.

Reid, Crowther Engineering designed the river pumphouse and pipelines. During 1968, Poole Construction Ltd. (PCL) built the pumphouse. During the early months of 1969, PCL laid steel pipelines on the ice on Yellowknife Bay to the Town and both mines. To reduce corrosion, the pipes had been internally and externally coated with epoxy.

They laid the pipes following the same procedure that had been used at Hay River. They laid the sections along the routes of the intended pipelines, and welded them together. Then they progressively cut the ice under the pipeline. At the same time they let air escape from a pipe slowly so that the pipe dropped into the lake slowly, beginning at the pumphouse and ending at the three delivery points.

* * * *

By 1964, Niven Lake had lost its summertime algae-green color for a wide area around the sewage inlet. It had become a gray color. Also, a twenty-centimeters-high island of sludge had developed near the sewage inlet. This part of the lake, only two hundred meters from a trailer park, was beginning to stink. However, if a deep hole could be excavated to trap the sludge, the odor would be reduced. Unfortunately, there was no dragline in Yellowknife to excavate the hole.

Figure 139. *Yellowknife Sewage Pond before explosion. Photo by Jack Grainge (1963).*

The public works crew and I put our heads together to solve the problem. This turned out to be a case where a few heads were not better than one. One of the crew had previously been a hard-rock driller at Giant Mine. He claimed that he could dynamite a deep hole in the bottom of the lake. He would drill and set the charges by the end of the day. I did not sleep well that night. I kept wondering how an ex-hard-rock miner had induced me to dynamite soft sewage sludge.

The next morning the miner, Stu Demelt, Glen Cinnamon, and I gathered around the proposed blast site. I worried even more when I saw the ends of about twenty-five three-meter-long rods sticking up all over the island of sludge. I wondered how the shallow dynamite could create a deep hole. The driller maintained the dynamite *would make a deep hole.*

The crew members took refuge in abandoned oil barrels that were lying on their sides near the shoreline. I stood on high ground about fifty meters away with my camera ready. Then the sludge blew. It rose about thirty-five meters high and the plume drifted about forty meters downwind. It was then that I noticed that a slight breeze was blowing toward the nearby trailer park. The sludge made a layer of muck up to five centimeters thick. It stank horribly, a pigpen-like odor. The breeze was not strong enough to dilute the odor, but carried it gently toward the trailer park. The secretary-treasurer received a phone call from a lady in the trailer park, "How would you like to be served fecal matter for breakfast?" (Not her exact words.)

Figure 140. *Explosion at Yellowknife Lagoon. Photo by Jack Grainge (1963).*

The sludge plume was high, the stink was high, but it was not a high point in my career. As the years went by, the distance of the far flung dung increased five-fold. A resident housewife claimed it landed on white shirts on her clothes line.

The following year, a backhoe was available and Thiel hired the operator to excavate a deep hole at the sewer outlet. The algae-green color returned to the water in the lake. Nevertheless a larger lagoon was still necessary.

In all shallow lakes, dead algae sink to the bottom, eventually becoming a thick layer of muck and the lake becomes a slough. During the 1970s, the heavy flow of nutrient-rich sewage, caused this to occur at the lower end of Niven Lake. If the muck were excavated and allowed to decompose for two years, it would have made fertile soil for gardens, valuable in Yellowknife.

During the 1960s, DI&NA constructed a twenty-man, minimum-security prison near the north end of Kam Lake. In response to their request for advice, I suggested that they discharge the sewage to a small lagoon overflowing to the lake. The system was satisfactory.

* * * *

Because Niven Lake had become overloaded, I recommended to Thiel that the City's sewage (Yellowknife had become a city) should be diverted to Kam Lake. Together we decided that no one used Kam Lake because everyone regarded it as being too polluted with Con Mine's mill wastes and sewage. Also, for two decades, it had received run-off wastes from Bevan's twenty-five cows, later reduced to eleven cows. The sewage load from each cow was equivalent to that of twenty people. Later it received effluent from the sewage lagoon of a 25-inmate Yellowknife Correctional Centre. As a result of the sewage, Kam Lake was nutrient enriched. Consequently, the fish were fatter than those in other lakes in the vicinity, including Great Slave Lake.

Thiel and I agreed that, if desired, the sewage could be piped a long way down the lake. The two lakes and wetland downstream of Kam Lake would provide complete removal of all sewage bacteria.

If the sewage were discharged at the far end of Kam Lake and the wastes from Con Mine diverted to a different chain of lakes, then Grace Lake and Kam Lake together could become recreational lakes.

The City Council hired Stanley Associates Engineering Ltd. to give an opinion. Frank Dusel, a senior engineer in the Company, recommended an overall study of the sewer system.

After making the study, he proposed that the sewage be pumped directly to Kam Lake. He designed a force-main to the high point of the rise, discharging to a gravity trunk sewer. Gravity sewers from Forest Park, South Forest Park, Matonabee and Frame Lake subdivisions discharged into the gravity section of the pipeline. The Yellowknife Correctional Centre began discharging untreated sewage into this trunk sewer.

In 1973, Thiel died in a boating accident. Kraft succeeded him as Public Works Superintendent. DI&NA provided a million dollars for the construction of the sewer main that would run to Kam Lake. Kraft authorized Reid, Crowther and Partners to design the pump station and pipeline to discharge the sewage from the whole town into Kam Lake.

After this construction was completed and the gravity sewers were discharging into Kam Lake, an official of a federal office sent a letter to John Parker, the Assistant Commissioner of the Northwest Territories. The writer stated that federal policy required that a natural lake should not be used as a sewage lagoon. He added that Yellowknife, the capital city, should have a sewage treatment plant, with aeration and chlorination. Being the capital city, its sewage treatment plant should be a state-of-the-art showcase.

Mr. Parker replied that Kam Lake was nothing special, there being a surplus of lakes around Yellowknife. The City was short of land for expansion and Kam Lake was out of the way and suitable for a sewage lagoon. Because it had been receiving mill wastes and sewage from Con Mine, it had no recreational value.

At a City Council meeting, I reported that Kam Lake, without any construction, would be an ideal sewage lagoon, itself a state-of-the-art treatment system. If however they wanted it for a recreational lake, the sewage could be piped to the far end of the lake. I stated that sewage lagoons were more efficient than treatment plants. I said there were a thousand or so towns in the central and western states and provinces served by sewage lagoons. Only towns with insufficient room for lagoons used treatment plants. Although the government would pay for the construction of the sewage treatment plant, the City simply could not afford its operational costs.

Far from being a showcase, the sewage treatment plant would seldom operate satisfactorily and end up a white elephant. I stated that my half brother, Rupert Littke, had been the engineer in charge of Edmonton's modern, sewage treatment plant. With more than a hundred employees, he had difficulty keeping it operating even reasonably well. Odors were sometimes a problem.

I showed the Council several slides of sewage lagoons in Scandinavia and Alberta, including some for which engineers of our office either had designed or had been responsible. These served military bases, national parks and Indian reserves. These were slides that I had been using in lectures to graduating engineers' classes at several universities, including the University of Alaska, and to three successive, triannual environment engineers' three-day courses in Anchorage, Alaska. I stated that Kam Lake served many subdivisions by gravity and to go to any other lake or lagoon would require a pump station. Pump stations are costly to both operate and maintain. Engineers in Scandinavia, leaders in environment protection, plan communities so that the least number of pump stations are required.

The NWT Water Board held a public hearing to air all points of view. Mayor Henne pointed out that because of the arsenic in Kam Lake, no one used it for either fishing or recreation. He stated that although the Government might pay for a sewage treatment plant, the city could not afford to operate it.

Ben Van Hees, a clever, innovative engineer residing in Yellowknife, and Peter Lawson, a senior engineer, both employees of Reed, Crowther and Partners, represented the City. They pointed out the value of Kam Lake as a lagoon. A federal wildlife officer stated that the sewage had made Kam Lake and the two downstream lakes into a wildlife haven. Van Hees, an experienced engineer stated that if necessary he could construct a dam across the Lake at the far end, and discharge the sewage there. The dam would make almost the whole lake recreational. I did not attend the hearing. I had already stated my opinion to the City Council.

Soon thereafter the Water Board of the Northwest Territories disallowed the use of Kam Lake for a sewage lagoon. The only feasible alternative was the construction of a pump station and five kilometers of force-main discharging to Fiddler's Lake, south of the airport. The Water Board required embankments around the lake to increase its volume. The NWT government paid the construction costs, and the City paid for maintenance and operational expenses. A few years later, they added major additional embankments.

References:

Christensen, V. 1997. (Assistant Deputy Minister, G.N.W.T.). Personal communication.

Dusel, F. 1997. (Senior Engineer, Stanley Engineering Co. Ltd.). Personal communication.

Fielden, R. 1997. (Engineer, Reid Crowther and Partners, Edmonton). Personal communication.

Foster, K. 1997. (formerly president, G.C.G. Dillon Consulting Ltd., Edmonton). Personal communication.

Jackson, S. (ed.). 1997. *Yellowknife, N.W.T., An Illustrated History.*

Kraft, J. 1997. Personal communication.

Milburn, B. 1997. *Rehabilitation of Municipal Services in Yellowknife, N.W.T.* Yellowknife: Government of the Northwest Territories.

Miller, F. 1997. Personal communication.

Miller, G. 1997. Personal communication.

Picard, J. 1997. Personal communication.

Price, R. 1967. Yellowknife. Toronto: Ray Price Sixex & Schuster of Canada Ltd.

Ward, M. 1991. *The Max Ward Story.* Toronto: McClelland and Steward Inc..

ARSENIC IN THE ENVIRONMENT OF YELLOWKNIFE

During 1948 and 1949, Con Mine had been discharging dark clouds of arsenic-laden fly ash into the atmosphere, without benefit of a high stack to scatter it. Consequently, the fly ash lay on the ground and snow in Yellowknife. On the snow it was thick enough to be seen as a pale gray dust. A Dene family living on Latham Island had been melting snow for drinking water. Their baby died from arsenic poisoning. Incidentally, the arsenic is in the form of arsenic trioxide.

Con Mine paid compensation to the parents of the baby. The owner of a dairy claimed his eleven cows had died from drinking water from the arsenic-contaminated Kam Lake. He received high compensation. He butchered the cattle and sold the meat door to door in Yellowknife. His replacement cows drank the water from Kam Lake, but remained healthy and produced milk, which was bought by people in Yellowknife.

Figure 141. *Con Mine arsenic dump. Photo by Jack Grainge (1963).*

Dr. Kingsley-Kay, Director, Industrial Health Services, Department of National Health and Welfare (DNH&W), and two of his chemists investigated the problem on the spot. Dr. Kay ordered the three mines, Con, Negus and Giant, to filter as much fly ash as possible from their roaster emissions, and construct high stacks to scatter the remainder over many kilometers around. For several years Kay's chemists checked the arsenic content of the stack emissions, in the ambient air, in garden vegetables and the arsenic settling in the countryside within and surrounding Yellowknife.

Disposal of the arsenic trioxide dust from Giant Mine's filters for gaseous discharges was a problem. Storage in pits would have resulted in rain dissolving them and washing them into Yellowknife Bay, everybody's drinking water source. The safest solution was to store it in large sealed excavations from which the ore had been mined. They used the excavations that were within a section of the permafrost, which in that area is 75m deep. They claimed that they would cement the arsenic in, if and when the mine stopped operating.

I thought about what would happen if the permafrost should melt. The melting process would start at the bottom. The mine is underlain by solid rock; however, there might be a few interconnecting cracks leading outward. Considering the huge flow of water in the Yellowknife River, this small seepage of arsenic would be inconsequential.

The committee appointed by the Canadian Public Health Association thought that Giant Mine's storage of arsenic trioxide underground was satisfactory. They stated in their report that Con Mine should have stored their arsenic wastes underground instead of storing them in a pond. At that time, Con was no longer mining ore that contained arsenic, and the pond had dried up.

Figure 142. *Growing plants (mushrooms) on arsenic wastes at Con Mine. Photo by Jack Grainge (1963).*

When I was there, they were covering the wastes with soil, and experimenting with methods of making grasses grow on it. A year or two later, they were digging up the arsenic trioxide and trucking it to a steel refinery in the United States.

To be on the safe side, successive chief medical health officers arranged a series of medical health surveys–Dr. D.L. Henderson in 1951-52, Dr. A.J. deVilliers and P.M. Baker in 1966, and Dr. O. Schaefer in 1975. Dr. deVilliers is an experienced toxicologist, while Dr. Schaefer is internationally acclaimed for his medical research in the North. After every survey the doctors reported that there was no indication that arsenic in the environment was affecting the health of the people. How much more reassuring it would have been if they had stated that the arsenic was not harming anyone.

In 1962, Dr. Otto Rath, Regional Director, Indian and Northern Health Services, DNH&W, asked me to prepare a report on the engineering aspects of the problem. My staff and I made the study with advice from Dr. T.J. Shnitka, Professor of Pathology, University of Alberta. The U.S. Public Health Service and the Health Department of California referred technical papers to me.

Dr. Rath and I concluded that arsenic pollution was under control. With few exceptions, the arsenic in the atmosphere and the water met US public health standards. At that time Canada had no standards. Arsenic concentration in the air was too low to cause any medical problem. The arsenic in locally grown vegetables, except lettuce, always met the standard. The weight of lettuce consumed by the people being small, the arsenic intake from that source would have been inconsequential. The arsenic in the soil was high, but of minor significance. Many towns throughout the world had much higher concentrations without noticeable consequences.

Nevertheless we continued to apply pressure on Giant Mine to reduce the emission of arsenic from their stack. After all, arsenic is a poison. They had been using electrostatic filters, so they added bag filters.

Our studies showed that most of the arsenic which people would assimilate, would enter their bodies by way of the drinking water. Using dye tracers, my staff and I studied the water currents in the Bay, but failed to find an arsenic-free channel from which to pump water. The Yellowknife River, which is upstream of the Bay, was the closest source of arsenic-free water.

The arsenic in the drinking water was high. Nevertheless, in our 1962 report, we did not recommend piping arsenic-free water from the Yellowknife River. Dr. Rath and I considered that in view of the low price of gold at the time, to do so might be too costly for the mines to bear. There was not a public health problem. It was just a question of obtaining water that was free of a poison, however small.

My report was widely circulated and no one suggested that a pipeline was necessary. Nevertheless in 1967, the Department of Indian and Northern Affairs (DI&NA) used it as the basis for ordering the two mines to each pay one third of the cost of a pipeline to extend from the water intakes of the town and both mines upstream to the Yellowknife River. I had to leave my wife at Expo in Montreal while I accompanied a DI&NA official to interview the senior executives of Falconbridge in Toronto for Giant Mine and COMINCO in Trail, B.C. for Con Mine.

* * * *

In 1971 Professor J.J. O'Toole, Iowa State University and three of his students made an educational field study in Yellowknife. They took samples of soils, water, air and some clippings of hair from a few people. They analysed these for arsenic content. They were proper, scientific sampling programs. In effect O'Toole's team repeated the initial tests that the chemists of DNH&W had made in 1949, before all of the corrective measures had been made. Professor O'Toole wrote a letter to Mayor Fred Henne telling him that the arsenic levels in these samples were dangerously high.

Mr. Henne became alarmed and reported the problem to the territorial and federal offices in Yellowknife. They set up an ad hoc committee, consisting of a dozen representatives of several offices in Yellowknife. They asked me to speak to them. At the meeting I stated that on the basis of our studies in1962, doctors in my department and I saw no cause for concern. Of course mill workers should follow rules to protect themselves from exposure to chemicals, but the public was safe.

To be certain that the scientists in the USA Environmental Protection Agency had not changed their opinion about the safe limits of arsenic, I telephoned Washington, D.C. and spoke to the head of the section responsible for these regulations. He said they had recently reviewed the subject and found their existing regulations to be satisfactory. The arsenic in the air and garden vegetables in Yellowknife met their regulations and the drinking water was free of pollution. There was no problem. The matter was settled.

However in 1975, Lloyd Tataryn produced an alarming segment of As It Happens, a Canada-wide, CBC radio program. He designed the program to show problems in Yellowknife's environment that were worse than those that existed in 1949, before corrective measures had been taken. He was an official of the United Steel Workers of America, which union had been turfed by the workers of Giant Mine in favor of a rival union. Obviously he was eager to show his union to advantage.

On the program, Barbara Frum telephoned Professor T. Hutchinson, Department of Environmental Studies, University of Toronto and Professor B. Carno, School of Public Health, University of Illinois. In answer to questions, those professors said Yellowknifers were in grave danger. Professor Carno claimed the water they were drinking was not even fit for swimming.

Having had no time to think about the questions, the professors had made understandable errors. I can recall many times when answering questions during my lectures, I have made some grave errors. Probably Hutchinson and Carno thought that the concentrations described were stronger than they were. Possibly on the spur of the moment, they were thinking that arsenic was a heavy metal. The atomic weight of arsenic is two-thirds that of cadmium and half that of lead and mercury. It is not even a metal. Thus they might have drawn inapplicable conclusions. For example, that arsenic ingested by Yellowknifers could not be excreted. It would accumulate permanently in soft tissues. Also arsenic would be ingested by lower forms of life in the Bay and ascend the food chain to the fish.

Arsenic is an in-between element. It can form both acidic and basic compounds. In pure form it looks like a grey metal. It is sometimes referred to as a metalloid.

The human body excretes arsenic slowly. Furthermore, long-time Yellowknife residents, who had lived there when arsenic in the air and drinking water was many times greater, had not contracted arsenic-related sicknesses. Obviously, their excretion rate equals their intake rate. Later, biologists tested fish taken from Kam Lake, in which the water contained ten times the concentration of arsenic as that in Yellowknife Bay. The fish were fat and healthy, and their flesh contained negligible concentrations of arsenic.

Professor Carno must have presumed that the arsenic concentration of the water in the Bay was much stronger than it was. Residents had been drinking water from the Bay prior to 1969, without suffering ill effects. People throughout the world, without harming their mouths, drink water from wells and springs that contain much higher concentrations of arsenic than the water in the Bay. Furthermore, the oral medicine, Fowler's solution, is ten thousand times more concentrated with arsenic than the water in the Bay.

The reports in newspapers were devastating. The Honorable Marc Lalonde, Minister of National Health and Welfare, immediately reviewed the studies by his department. Dr. D.L. Henderson in 1951-52, Dr. A.J. deVilliers and P.M. Baker in 1966 and Dr. O. Schaefer in 1975, all of which concluded that there was no danger to the public health. In a news release, Lalonde then stated that there was no health hazard to Yellowknife residents. Unfortunately, some people were not convinced, so the newspapers did not let the matter drop.

* * * *

From 1975 on, engineers and biologists in the federal service made studies of arsenic in the environment similar to those made in 1949 by the chemists in DNH&W, none of which were related to health effects. In 1976 the previously mentioned committee in Yellowknife asked me to prepare a report combining all pertinent information. I complied under the supervision of Dr. T. Covill, Chief Medical Health Officer and his assistant, Dr. Richard Eaton, later succeeded by Dr. E. Fischer.

A supervisor in another department decided they would submit a separate report. Dr. Fischer and I visited the supervisor of the engineer in the other department. Dr. Fischer pleaded with him not to submit a separate report. He rightly predicted that the press would headline a dispute between two civil servants. It is a great way to sell newspapers.

The Yellowknife newspaper said my submission was "easy to read." I had stated in my front summary that since 1969, when arsenic-free water was pumped into the Yellowknife water supply from upstream of all sources of arsenic pollution, there was no danger of arsenic in Yellowknife's environment causing health problems. Also nearby wildlife and water life were healthy.

My report included summaries of fifty-two scientific reports, many concerning health effects resulting from various levels of exposure to arsenic in the air, water, food and skin contact. These studies had been conducted in cities and towns throughout the world. The other report consisted of summaries of the reports of concentrations of arsenic in the air, water, vegetables grown locally, puddles, dirt, and grass in the vicinity of Yellowknife.

Newspapers made much of two civil servants not being able to prepare a single report. Therefore the Honorable Marc Lalonde asked the Canadian Public Health Association to appoint a task force, independent of the government service to prepare a report.

The task force consisted of two medical doctors and a provincial sanitary engineer. They called a public meeting in Ottawa. Dr. Schaefer and I were asked to sit at a side table near the front. I reported the problems my predecessors, my engineers and I had faced, and that we had corrected them.

In 1949-50, roasters of three mines, Con, Negus and Giant, were spewing unfiltered arsenic-laden fly ash into the Yellowknife's atmosphere. At the time of the meeting only Giant Mine operated a roaster. Most of the fly ash from their roaster was filtered out and most of the emissions were scattered beyond Yellowknife by means of a high stack. The water distributed in the water systems of the City and the mine communities were coming from Yellowknife River, upstream of all sources of pollution. I regret that I did not review in more detail the work we had done. I should have done so, but I did not know that I would get no opportunity to do so later.

Figure 143. *Aerial view of Yellowknife. Photo by Jack Grainge.*

Lloyd Tataryn, the clever labor union organizer, stood in front of us badgering us with questions that he repeated quickly. I could see how he was able to get two scientists on the

As It Happens radio program to make ridiculous statements about the water in Yellowknife Bay not being safe for swimming. I did not understand why a labor union organizer should have apparently been running the public meeting. Dr. Schaefer answered back. My short-term memory failed and I was slow in answering a simple question about what part of the lungs was harmed by the arsenic in the air. One section of my report by a medical doctor dealt specifically with this subject. In this I had quoted her as saying that some large particulate would stick to the walls of the passageways and some small particulate would reach the sacs at the ends. The mine filters had removed the large particulate. Most, but not all of the inhaled small particulates would be exhaled by the person breathing them in. I finally answered, "The sacs."

I did not get another opportunity to speak. Tataryn had demonstrated that by my hesitation I showed that I was not a knowledgeable witness.

The Task Force Chairman asked Dr. Covill to send all of our information regarding the studies of the arsenic in Yellowknife's environment. I sent them a carton of the studies of arsenic in the environment in various places throughout the world on which I had based much of my report. I suppose the medical doctors based their report on the four, previous medical studies that had been made. They agreed with the conclusions of all the medical surveys of the people in Yellowknife. These studies had shown that there was no evidence that people's health had suffered from exposure to arsenic.

In its report the Task Force concluded that the water was safe to drink, the air safe to breathe, washed garden vegetables and berries safe to eat, and that the arsenic in the soil posed no health problem.

A few years later when improved bag filters were obtainable, Giant Mine installed them. Also much of the buried arsenic dust at Con Mine was dug up and trucked to United States for reprocessing. Gold was extracted from the dust and the remaining arsenic used in the manufacture of steel. Nevertheless mining engineers rightly continued to look for new methods of reducing the arsenic emmissions to the atmosphere.

In 1949 the Yellowknife residents were breathing air that was badly polluted with arsenic-laden fly ash. A few people were drinking water melted from snow that was thick with arsenic. With the exception of a Dene baby who had consumed heavily contaminated snow-melt water, the residents had shown no signs of arsenic-related sicknesses.

Since 1969, the residents have been enjoying the purest of water and clear skies, and they continue to do so.

Reference:

Bromley, B. 1963. (Long time resident of Yellowknife). Personal communication.
Schaefer, O. 1963. Personal communication.

RAE-EDZO

In 1790 the North West Co. established Old Fort Providence approximately fifty kilometers southeast of present Rae, and closed it thirty-three years later. In 1852 the HBC established Old Fort Rae on Fort Island, in North Arm of Great Slave Lake, 24 km south of present Rae. They named it after Dr. John Rae, the chief factor of the Mackenzie District, stationed at Fort Simpson. Rae was famous for his arctic exploration. He found the remains of the Sir John Franklin expedition at Victory Point on the northeast shore of King William Island.

In 1859, Father Grollier visited Old Fort Rae. In 1872, Father Roure built a log-cabin there and six years later, St. Michael's Mission. Fifteen years later Hyslop and Nagle established a trading post at the present site of Rae. In about 1905 the HBC and a Catholic mission were built at the new site. The Catholic Brothers constructed Faraud Hospital in 1936-40, and ten Sisters of the Grey Nuns staffed it. The Department of National Health and Welfare (DNH&W) provided a resident doctor. In 1948 the Department of Indian and Northern Affairs (DI&NA) constructed a school and a teacher's house. In the fifties, the RCMP established a one-man post. The settlement served the Dogrib Dene who occupied a wide tract of land extending from Great Slave Lake to Great Bear Lake.

Figure 144. *Aerial view of Rae, showing bare rock–the bane of the early settlement, 1959. Photo by Jack Grainge.*

In 1931-32, Ab Coyne, now a retired, successful businessman in Edmonton, was working for one of the four trading posts at Rae. His post was situated on an island in Marian Lake, or perhaps another lake. On his first day he and Peter Erasmus, their interpreter, canoed to the mainland, where they cut down trees, 'delimbed' and piled them for firewood. They left the firewood there to be hauled home by dog sled the following winter. The first day Coyne worked hard for twelve hours. He needed to take a bath, and five months later he did so.

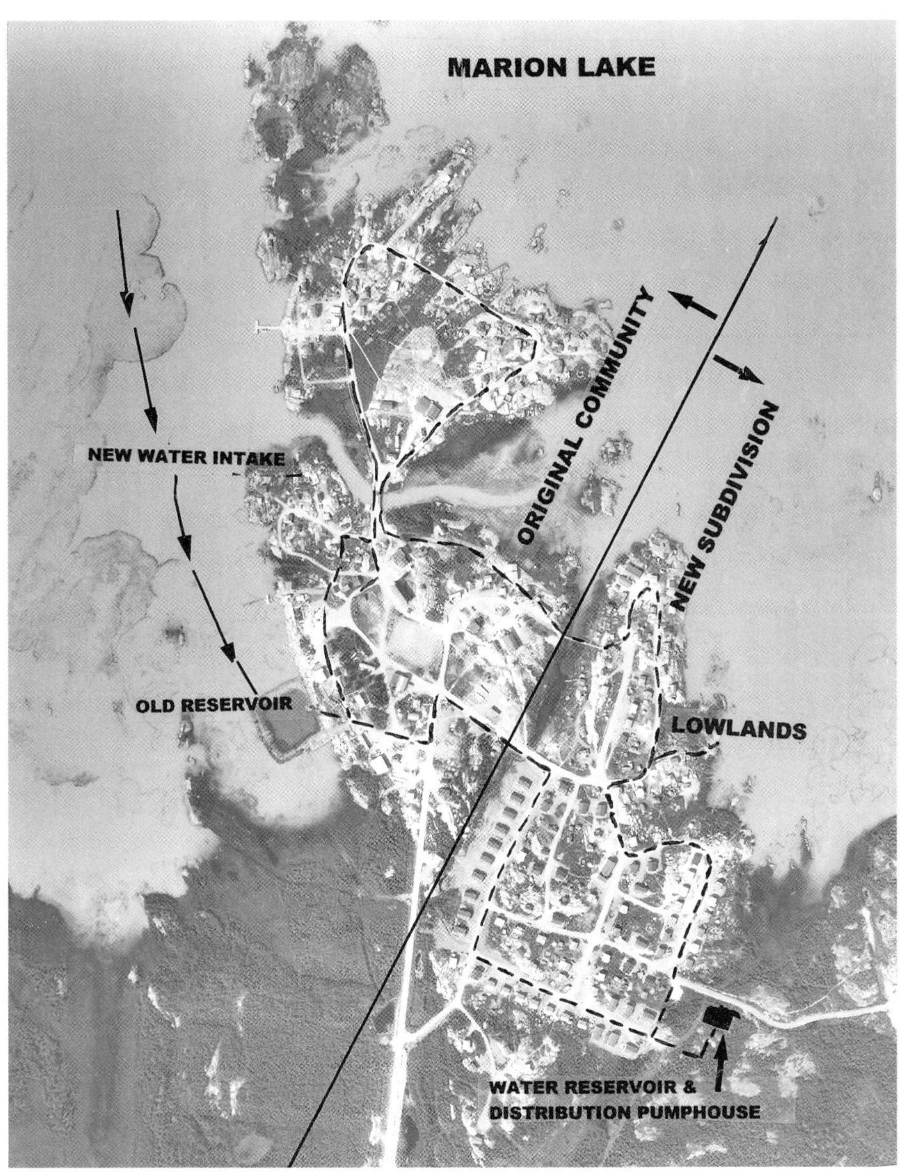

Figure 145. *Aerial photo of Rae. Department of Energy, Mines, and Resources, Canada, 1979.*

Coyne was the water boy. With an ice chisel, he chipped a hole through a meter of ice. The hole was cone-shaped, a meter wide at the surface. That was hard work. I suppose pointed, triangular cross-section, needle bars had not been invented then. Using a needle bar with all three edges sharp, one can cut a thirty centimeter diameter hole through thick ice in one third the time.

Coyne told me that most of the Dene hunted and trapped in the near North, returning to Rae at Christmas, Easter and treaty time. Many of the Dene bought boats in Rae and paddled them up Snare River and lakes to the tundra region. The dogs ran along the shore, keeping abreast of them and joining them for feed when the people came to shore to camp. After two winters of trapping, the Dene would abandon their boats and return to Rae, their sleds loaded with white fox pelts.

Figure 146. *Log houses. Photo by Jack Grainge.*

At that time Rae was possibly the busiest community in the NWT as it was the outfitting place for prospectors heading north, mostly to the area east of Great Bear Lake. Most of them arrived by plane and bought boats and supplies for traveling to prospect sites. Young Coyne was excited to become acquainted with two of the prospectors, the Russian Prince Leo Galitzime of Russia and hockey-great Lionel Conacher.

From the turn of the century on there were severe outbreaks of measles, tuberculosis, influenza and gastrointestinal diseases among the Dene in Rae. During the 1918 influenza epidemic, half of the 4000 Dogribs died. Probably these epidemics raged because large numbers of prospectors with low levels of these diseases passed up the Snare River, the hunting grounds of large numbers of Dene. The Dene had almost no immunity to these diseases.

* * * *

In 1952 Stan Copp and I accompanied a friend flying to Rae. We visited the resident doctor and looked around the community. In every direction I saw beautiful scenes. I observed the white-colored hospital, mission, HBC, RCMP, school, doctor's and teacher's houses and a few tents, all on huge, bulging bare rocks. A few boats and our plane were tied up along the shores of a calm, mirror-like lake almost surrounding the community. There were green forests in

Figure 147. *Shrine. Photo by Jack Grainge.*

the distance. The scenery reminded me of photos of picturesque fishing villages in Newfoundland. Even today I can close my eyes and see those exquisite scenes.

The lake is shallow. Consequently wind-caused currents stir up the bottom silt, rendering the water too turbid to drink. Some people would let it settle in jars for a couple of days. The doctor pumped water from the lake into a water reservoir below his residence. When he started the pump, he poured some alum into the tank. The inrushing water mixed the alum and water. After the pumping the alum caused the particles to settle quickly.

Unfortunately, the next time the reservoirs were filled, the incoming water stirred up the previously settled silt. Thus, for a short while the water in the tank was more turbid than the water being pumped from the lake.

Figure 148. *Summer tent camp of Chief Bruneau on shore of Russel Lake. Photo by Jack Grainge.*

The water supply system serving the Grey-Nun's hospital, and R.C. mission consisted of a suction pipe extending underwater forty meters into the lake, a pump and a water reservoir in the basement. The far end of the water-intake pipe was turned up to avoid sucking bottom sediment. One winter the ice froze deep, plugging the far end of the pipe.

The teacher, the RCMP and the HBC manager carried water from a nearby channel in which the water was fairly still, allowing some of the sediment to settle out. In winter they chipped holes through the ice in the lake.

The hospital, mission, RCMP and doctor's house contained sewage holding tanks. A hospital employee pumped the wastes from the hospital, mission and doctors residence into a tank on a tractor-drawn stoneboat. He hauled the sewage over a water-soaked road to some bush land. Other outsiders emptied their honey buckets in that direction.

* * * *

On a sunny summer day in 1956, the ever-joking, perceptive Dr. L.E.C. (Bill) Davies and I, together with the much-concerned local doctor, tramped around Rae. The buildings and tents were on huge, bare, granite rock mounds, interspersed with patches of muskeg. There being no soil to seep into, wash water and body wastes could not seep away. They flowed in ditches past houses and then, along with dog manure, into the lake–everyone's water source. Obviously visits by outsiders with diarrhea caused outbreaks.

In my report I detailed criticisms of the situation. I made a few recommendations for hauling sewage a distance away, but later realized that my ideas were not practical.

I returned to Rae in 1959, accompanied by Jim Anderegg and LaMar Hubbs, two American engineers conducting research in my field in Alaska. At the time DI&NA was planning to construct a power plant, a large school and a costly twenty kilometer road connecting the community to the highway from Yellowknife to Alberta, which was then under construction. The

Figure 149. *Ethel Martens was the first instructor of nurses and other public health officials. Photo by Jack Grainge.*

three of us discussed the sanitation problems of the Dene with the local doctor, priest, teacher, HBC manager and the RCMP constable, and later in Fort Smith with Curt Merrill, the Territorial Administrator, Ken Hawkins, Chief Engineer and Bishop Fallaize. We pointed out the difficulty of providing sanitary conditions at the present site. The indiscriminate bulges of bare granite rock made future, subsurface piped water and sewer systems impossible.

All agreed that sanitation conditions in Rae were dreadful. On the esker on the south side of the lake, there would likely be a suitable site for a new community. We thought that Rae itself should remain only a small fishing village. It would be a situation similar to the construction in Yellowknife of New Town alongside Old Town. Sanitary conditions in Rae were worse than those in both Old Town Yellowknife and Aklavik. Also the incidence of intestinal diseases and deaths from those diseases in Rae was alarmingly high.

Figure 150. *Dr. Sullivan with Chief Bruneau to left and medicine man with cane to the right. During his days as Chief, Bruneau kept order in the community–his word was law. Photo by Jack Grainge.*

We also agreed that before building a proposed new school, a large power plant and a substantial road into Rae, senior officials in Ottawa should decide to move Rae. We could not discuss the matter with the Dene because Chief Jimmy Bruneau was not in Rae. He had become chief in 1936, and had proved himself to be a respected, wise man. He died in 1975.

If Ben Sivertz, Commissioner of the Northwest Territories in Ottawa, saw our recommendation, he must have rejected it. I understood that he did not discuss the matter with anyone in DNH&W or even Curt Merrill, District Administrator in Fort Smith.

Figure 151. *The rocky terrain made construction and servicing especially difficult. Photo by Jack Grainge.*

Figure 152. *Rocky terrain in Rae. Photo by Jack Grainge.*

With the directors and chief engineers in Ottawa, the consulting engineer in Edmonton and the district engineer in Fort Smith, mix-ups were not surprising. At the same time an Edmonton engineer was planning a bridge over the several channels connecting Marian Lake with Great Slave Lake. The vessels use only the east channel, which is the widest and deepest. The bridge was built with the high end over the west channel. When the planners for Rae are in Edmonton, Fort Smith and Ottawa, best-laid schemes gang aft a-gley.

After the much higher than expected costs of constructing Inuvik, moving communities may have been taboo among planners. The Inuvik cost was too high, about $25 million. The main reasons for the high costs at Inuvik had been the permafrost and the costly, high, above ground utilidors serving widespread buildings. Probably the planners in Ottawa had been pleased with the success of the creation of New Town at Yellowknife, at a cost of less than a million dollars, and did not realize until too late the enormity of the final bill for Inuvik. Of course hindsight is great.

The difficulty was the remoteness of the Ottawa officials from the problem. If the decision makers could have seen the site of Rae, they likely would have agreed. Even the engineers in Yellowknife had to take a plane to get there. I had no difficulty with DIA&NA officials in Alberta. I could speak directly to them, and they could drive to see the health problems on Indian reserves and in national parks. Therefore they could take faster, more appropriate action.

Figure 153. *Fr. Duchaussois boating author to Chief Bruneau's summer camp. Photo by Jack Grainge.*

Figure 154. *Chinking log cabin with mud. Photo by Jack Grainge.*

At Rae the high cost of future construction of infrastructure on extensive rock billows could be avoided by relocating the community on the firm soil of the wide esker on which the Edmonton-Yellowknife highway was to run.

In 1961, long after the power plant and access road from the highway had been constructed. I discussed the problem with a regional administrator stationed in Yellowknife. He said that Chief Bruneau opposed moving the settlement. I therefore drove to Rae. Father Duchaussois boated me to Bruneau's fishing camp, on the shore of nearby Russell Lake. Bruneau said that he would not object to the settlement moving south of Marian Lake, but that he himself would remain in Rae. He wanted to be near his boat.

Unfortunately many influential people in Ottawa and Yellowknife believed that the strong Chief Bruneau opposed moving the community. It was impossible for me to explain what I had heard to all the widely located officials involved.

A month following my 1959 examination of Rae, John Kerr, a new engineer in our office visited several communities. While in Rae he sent a two-page wire to Prime Minister John Diefenbaker describing the deplorable, unsanitary conditions in Rae. Believe me, I heard about it.

* * * *

C.C. Parker Engineering Ltd., of which Dean Whitaker was the manager of the Edmonton office, prepared plans for the construction of a water supply system for Rae. Ken Ford, Project Engineer, asked me for suggestions. I explained the problems resulting from the lake being shallow, the water being turbid and lack of soil cover in the settlement. I stated that because there was no soil cover on the rocks, wastewater could not seep away through soil. Runoff water containing sewage polluted the water supplies of the Dene. Privies with pits were impossible. A system to provide safe water for the Dene in Rae would be prohibitively costly.

Ford designed a water supply system. The water source would be a 60 m by 60 m pond within the lake. The embankments would be built with silt and clay excavated with a dragline from the middle of the reservoir. Water from the lake would flow into the reservoir through pipes under the embankment. The pond would be deep, so that water could be pumped from it throughout the winter.

The water would be piped to an above-ground reservoir in a heated building in which it would be chlorinated. Water would be piped through an above-ground utilidor to the school, RCMP, hospital and a hydrant beside the hospital for the Dene. Because the hydrant would be much farther from the Dene homes than the lake shore, many of the Dene would not use it.

In a phone call to Dr. Falconer, Regional Director of Medical Services Directorate, DNH&W, I criticized the design. Dean Whitaker invited me to see his new office and share a coffee. He and I together with our wives were bridge playing friends. Discussions turned to Rae, and he asked why I had found fault with their design. I described the rolling granite topography, and said that the Dene problems were being overlooked. I said that the Mackenzie Highway was built on a large esker and that there must be sites there where practical, subsurface water mains and sewers could be constructed. He told me that it sounded like the plans did not address the long range problems. He said he

Figure 155. *Council meeting; Chief Bruneau with pipe. Photo by Jack Grainge.*

would discuss the matter with the DI&NA engineers who had hired him. I do not know what was discussed, but there were no changes in the design.

After the construction of the water system, outbreaks of intestinal diseases continued. After all, people continued to obtain contaminated water from the lake. Drs. Ken Butler and Tom Orford, DNH&W, Edmonton, and I met with Curt Merrill, Regional Director, and Ken Hawkins, Regional Engineer, in Fort Smith. We stated that poor engineering had resulted in high incidences of sickness and deaths among the Dene in Rae. Merrill and Hawkins made no suggestions that they would be trying to convince the Ottawa decision makers. I believe that they realized that the decision makers in Ottawa had already considered that alternative and would not consider discussing the matter again. Usually Merrill and Hawkins were most cooperative with me. If we could have made a site examination all together, the doctors and I might have been more convincing.

In 1965, Mike Hewitt, Region Public Health Nurse, later a NWT deputy minister in Yellowknife, asked me to attend a Dogrib Council meeting in Rae. At the meeting Hewitt reported the history of frequent disease outbreaks and consequent deaths. One of the elders suggested that they hold a meeting of all the people that evening. At that meeting Hewitt again reported the sicknesses and deaths. Several people stated that they knew about the sicknesses and deaths but no one had explained what had caused them. They agreed that we should look for a better site. Ed Zoe, a most enthusiastic fellow, offered to help us.

We found sandy, flat land northwest of Rae. We considered the site large enough but the land was low and flat. It would be impractical to lay shallowly buried pipes through which the sewage would flow by gravity. The sewers would either become too deep at the ends of streets or require an excessive number of costly sewage pump stations.

We next found a sandy site south of the Mackenzie Highway. It was much larger than the one near Rae. It had the advantage of being sloped so that shallow, buried sewers could drain by gravity. Being shallow, they would be easy to repair, especially in winter. If they used a long intake pipeline, Great Slave Lake would be a source of clear, good-quality drinking water that would require no treatment other than chlorination. There could be waterfront lots on Great Slave Lake. Close access to the Mackenzie Highway would favor the construction of a hotel, restaurants, service stations, stores, shops for handicraft sales and boat sales and rentals.

Figure 156. *Ed Zoe at one possible new site along highway. Photo by Jack Grainge.*

We had proved that a better site was available. Since the main highway was on an esker, most likely there would be better sites alongside of it.

In 1966, Norm Lawrence, President, AESL, found an extensive, gently sloping site between the highway and the south shore of Marian Lake. The soil was sandy, interspersed with small patches of silt. The site was large. There would have been room for playgrounds, sports fields and a small airport. It could have waterfront sites on Marian Lake and West Channel and a short access road to Great Slave Lake.

Being near the Alberta-Yellowknife highway, the people could construct a store to sell native art and a motel with a restaurant and service station to serve travelers.

* * * *

Don Gamble, AESL Project Engineer and Jim Cameron, his assistant, both young, imaginative and capable engineers, managed the construction of the water and sewerage systems. Initially a well was drilled, but the water was salty. Cameron and I boated out from the shore of Great Slave Lake to see if clear, soft water was available within a reasonable distance of the shore. I forget how far it was but the design engineers decided that West Channel was much closer and the water was suitable.

The water system is standard. A coated steel intake pipe extends fifteen meters into West Channel, leading underground to two wet wells below a pumphouse. From these, water is pumped to a manufactured, automatic, Waterboy treatment plant. Treated water discharges into a subsurface, concrete, treated water reservoir.

According to the record the water is hard, unlike the water in Marian Lake. Yet it is Marian Lake water that flows through this channel on its way to Great Slave Lake. Also, if the analysis is correct, the water also contains much undesirable manganese, not typical of the surface water. I wonder if ground water is seeping into the wet wells.

Shallow systems of water mains and sewers are pre-insulated. Sewers discharge to a sewage pump station and then into two sewage lagoons on the opposite side of the Alberta-Yellowknife highway. The lagoons provide a total of more than a year's retention time and discharge to Great Slave Lake. There is adequate space for the construction of additional lagoons when required.

In 1969, surveying and some road construction for the new townsite began. In 1970 construction started on water and sewer systems and buildings, several houses, a six-room school and student dormitory. DNH&W built a nursing station. A private individual built a store.

Initially the people were enthusiastic about the proposed new townsite. They named it after their esteemed former Chief Edzo, who had arranged an advantageous peace treaty with Chief Akaitcho of the Yellowknives.

However the HBC refused to move their post without financial compensation. The people would not move away from the HBC post because it was a social meeting place. The DI&NA administrators were not as generous as the Honorable Jean Lesage had been when moving businesses and institutions from Aklavik to Inuvik. Lesage knew how to move a settlement fast, before people could conjure up objections to the move.

Dave Geddes, a retired teacher in Edmonton, was in the early 1950s a clerk at the HBC post in Rae. Recently he confirmed my opinion about the post being a social meeting place. He said a man would bring his furs to the back of the store and bargain, and then visit his friends in the front of the store. He might go away and return the next day to bargain and visit. When they finally reached a settlement they would bind the deal with a handshake.

Then they would go to the front of the store, where he would visit friends and buy perhaps a package of sugar. In the afternoon he and his wife would return, visit their friends and buy another small item. The next day they would be back again and so on.

Geddes told me that he and the manager admired the people for their modesty and the way they cared for one another. They were friendly and when they met after a long trek by dog team, they would shake hands. That hand shake said, "I'm glad you made it." The people were hard-working, successful trappers, hunters and fishermen. They were honest and would never touch anything on the HBC shelves without permission.

Later, Northern Health Services, a new division of DNH&W, became organized with the articulate Dr. Gordon Butler the director. He had had considerable education in public health and much experience in Africa. He quickly sized up the situation in Rae. He discussed the matter with Stuart Hodgson, Commissioner of the Northwest Territories, and pointed out the deficiencies in the old site to him. Dr. Butler and Mr. Hodgson became personal friends.

Without an HBC store and other social facilities, few Dene moved to Edzo. They preferred to stay in Rae even though their children had to be bussed to the school in Edzo.

Mr. Hodgson favored the move to Edzo. The Government of the NWT (GNWT) implemented an unofficial building freeze on Rae, and required all new houses to be built in Edzo, none in Rae. Not wanting to waste funds in case of an eventual move, the GNWT delayed the construction of a community hall in Rae.

The NWT Council considered the matter at a meeting in Yellowknife. Mr. Hodgson asked Gordon Gibson, a NWT Council member from Vancouver, Dr. Lloyd Barber, an NWT Council member and a professor of geography, University of Saskatchewan, Norm Lawrence and me to meet together after supper.

The NWT council member from Vancouver stated that DI&NA should haul good soil to cover Rae to a “depth of several feet.” The cost, however, would have been many millions and eventually rains would wash the soil off the smooth rocks into the lake. The lake would become even shallower than it already was.

Lawrence and I maintained that the site of Rae was hopeless and the one at Edzo was ideal. Dr. Barber said that he would be reporting to Mr. Hodgson. Mr. Hodgson remained no closer to making a decision. However the Dene continued to pressure him for a school, a community hall and houses in Rae. Eventually he agreed, lifting the building freeze on Rae.

In 1976, Dr. Tom Covill, Dr. Butler’s successor, asked me to go to Rae. He said that the people were unhappy about the unsanitary conditions. They were complaining that Health and Welfare officials had shown no concern. We thought that ironic.

I made an inspection of Rae. I examined the manufactured, Waterboy treatment plant and the treated water reservoir. The plant provided coagulation, mixing, settling, filtering and chlorination, and at regular intervals, automatic back washing of the filters. The local superintendent of Public Works deplored the excessive costs of development in Rae. He was the one who told me that the people would have moved to Edzo if the HBC had moved their store. That cost would have been minuscule compared to that being spent trying to make Rae livable.

There was a garbage collection system and a satisfactory disposal site in a large, deep pit between the road and Marian Lake. Sewage trucks dumped sewage into a shallow lagoon on the opposite side of the road.

Construction of the sewage lagoon entailed blasting an extremely costly, forty-meter long, deep drainage trench through solid precambrian rock from the lagoon to Marian Lake. Plastic toilet bags were thrown down a chute into the lagoon. The full bags floated around, and in the winter piled up on the ice. I expected that the toilet-bag chute would become mucked up within a year. It was not a satisfactory disposal. However I could think of no better method other than hauling the bags to a new, distant site off the Mackenzie Highway.

I talked to many different people. Dr. Green, dentist, and the public health nurse were disgusted with the Rae site. They said that the engineers were wasting money trying to make it habitable. I interviewed a tribal elder in his home near the lake. He told me that he wished the community would move to Edzo. However he and several other people with lakeside lots would not move. They would want to be near their boats, to protect them from vandals. This I agreed was reasonable. It is the same thing that Chief Bruneau told me in 1961.

The local administrator tried to convince me that Rae was a good site. After listening to me describe the unsanitary conditions being experienced by the Dene, he seemed less sure of himself. He later congratulated me for my report about the awful unsanitary conditions in the community.

In my report, I described sanitary conditions for the Dene being far from acceptable. I said that even if huge sums were spent in Rae, the community plan would remain a poor one. I recommended that steps be taken to expand Edzo but not Rae.

The following summer George Woodget, Region Environmental Health Officer, DNH&W, and I examined both settlements. We considered that if there were no restrictions, Rae would continue to grow. Water mains and sewers in above-ground utilidors would become a costly necessity. They would cut up the community including recreation fields. We thought that in time people would begin to move to Edzo.

References:

Sanger, M. ca. 1991. *The Community of Rae.* GNWT: Report of the Department of Indian and Northern Affairs.

C3. Catholic Archives in Yellowknife

Coyne, A. 1998. Personal communication.

Fuller, W.A. 1998. Personal communication, Professer Emeritus–Botany, University of Alberta.

Geddes, D. 1998. Personal communication.

Lawrence, N. 1998. Personal communication.

Schaefer, O. 1998. Personal communication

HAY RIVER

From ancient times, the mouth of the Hay River was most likely a meeting place for families of Dene. Traveling by canoes or dog sleds, people living along the shores of both the river and Great Slave Lake could meet.

In 1868, the Hudson's Bay Co. built a trading post on the east shore of East Channel, across the River from Vale Island. The following year, Father Gascon established a Catholic Mission and closed it six years later. In 1894, Rev. T.J. Marsh established an Anglican mission. The following April, Charles Camsell, later to become a federal deputy minister, arrived while delivering the mail by dog-team. He stayed a few weeks helping Marsh whipsaw lumber for a church. The population was 60 to 70.

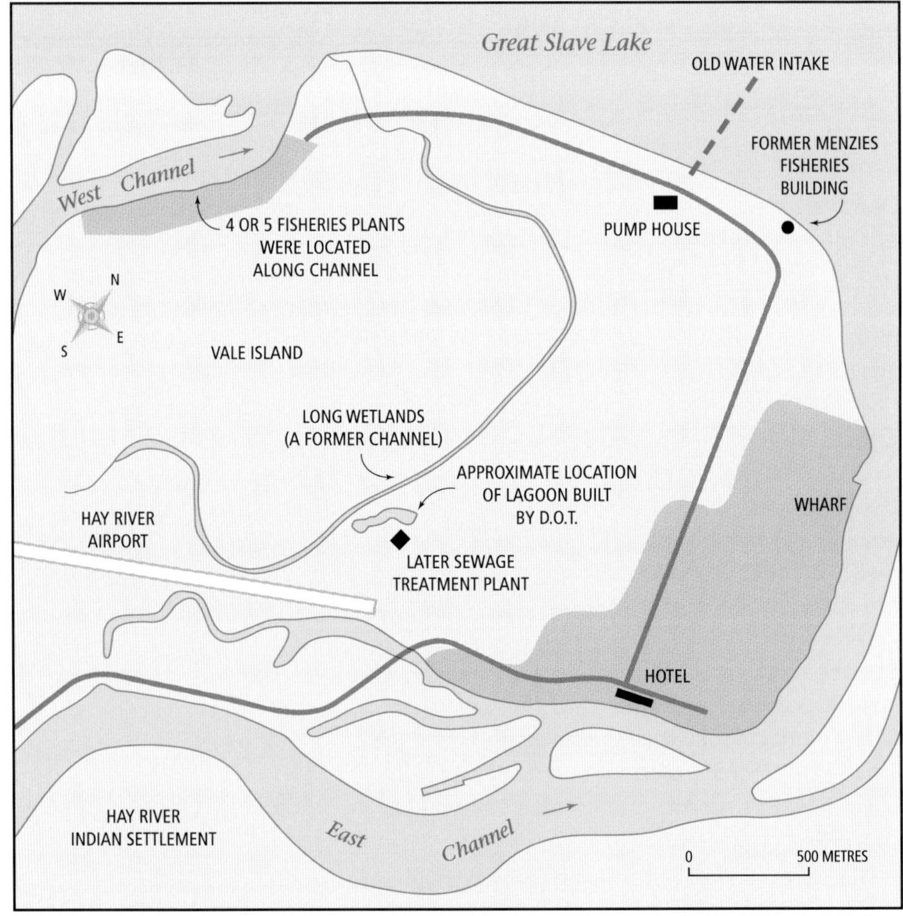

Figure 157. *Map of Hay River. Adapted from EPEC Consulting Western Ltd. (1981), by Johnson Cartographics Inc., Edmonton.*

In 1925 the RCMP established a detachment in Hay River. A decade later people began building houses on Vale Island, on the west shore of the East Channel. In 1942 the USA army built a gravel airport on the sand-silt island. This was one of a chain of airports they were building from Edmonton to Norman Wells. They required air transport of personnel for drilling oil wells at Norman Wells and laying an oil pipeline from there to an oil refinery they were constructing in Whitehorse.

Figure 158. *Freight vessels wintering at Hay River. Photo by Jack Grainge.*

At that time there was no road connection from the NWT to Alberta highways. However in the early 1940s, John Dennison, the indomitable, ice-road builder, confounded everyone by trucking along trails through the bush to Peace River. He was accompanied by a spare driver, Larry Scheck.

Following the war the federal government constructed a gravel road from the railway town, Grimshaw, Alberta to Hay River. That was the beginning of a large port

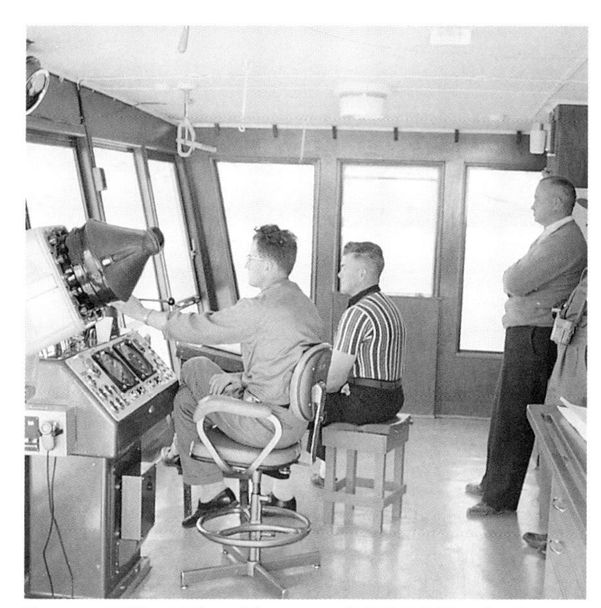

Figure 159. *Wheel house of CCGS Tembah. Photo by Jack Grainge.*

town on low-lying, Vale Island. The soil is silt and sand with ice-rich permafrost in some places.

In 1950 Ken Gaetz, an energetic Pentecostal missionary, built a church on the island. He also traveled by scow bringing his message to settlements along the Mackenzie River.

Responding to local pressure to build a hospital, in 1953, DNH&W constructed a nursing station on the island. Beatrice Purdy became the first public health nurse. Later she married Dr. Doug Abbey, the first local medical doctor.

Figure 160. *Launching buoys to guide freight vessels. Photo by Jack Grainge.*

Local public pressure for a hospital continued but DNH&W did not comply. Gaetz solicited funds from southern Penticostal Assemblies. The Stone Penticostal Church in Toronto used a bequest from the estate of Mr. H.H. Williams to build a hospital in Hay River. It opened in 1965 and was named after him.

While in Hay River, I happened to meet the NWT superintendent of the Canadian Coast Guard Service. He offered to transport me to Norman Wells on the CCGS Tembah. They set out buoys to mark routes along the rivers which were deep enough for the passage of freight vessels and barges.

* * * *

During the summer of 1951, Stan Copp, and I drove to Hay River. We found most of the people living in small log or wood-frame houses in a narrow strip of Vale Island, fronting both East Channel and the Lake. A few Dene lived in the old town on the mainland, on the opposite side of East Channel.

R.J. Douglas, the DI&NA district engineer, gave Stan and me a tour. A DI&NA employee hauled water from a pump house at the lakeshore in a tank truck. Some people had indoor plumbing with water reservoirs, septic tanks and seepage pits. Others had privies, and discarded wastewater near their houses.

Douglas told us he was going to replace the bridge across West Channel from the mainland to Vale Island with a causeway. In retrospect I realize that all three of us should have considered the implications of blocking a channel of a delta. A few months later Douglas left to take a job with Mannix Construction Co. in Calgary, but not before he had built the soon-to-become-infamous causeway.

Don Stewart, the Regional Supervisor, Department of Fisheries (DOF), showed us his office-residence, the only building in the community with a basement. The concrete basement floor had become water-soaked, heaved up and cracked into fifty pieces. We thought that the causes were poor concrete and uneven settlement of the subsoil. Later we realized that permafrost below the floor had melted and settled. Stewart was later to become a partner with Lionel Gagnier in a contracting company and hardware store. Later he became mayor of the town and eventually speaker of the NWT Council.

In approximately 1948 Ed Berg built a long, two-story, wood-frame hotel. In approximately 1954, I stayed there. Settlement of the foundation at the midpoint had resulted in the building becoming swaybacked. I could not close the toilet door, even though someone had sawed a 10 cm, wedge-shaped piece off its bottom. That cut-off might have allowed the door to close for a while, but not when I was there. The open door shielded toilet users from the direct view of people in the hall. The tub was not shielded, so I remained grimy and unwashed.

I was grimy because the silt on all the lightly graveled roads had worked to the surface and become wind-blown everywhere. The cheerful, friendly, Dene hotel maids worked hard to keep the hotel clean, but they were only partly successful.

* * * *

During the summer the swamp-colored water flowing from Hay River joins the similarly colored water from the Slave, Sandy and Buffalo Rivers. Hugging the shore, water from these rivers drifts westward toward the Mackenzie River outlet of Great Slave Lake. During annual spring floods, the colored water under the ice extends more than six kilometers from the shore.

In 1951 the water pump was located in a house on the shore of Great Slave Lake, five hundred meters west of the mouth of the river. The system consisted of a galvanized iron, buried intake pipe, a pump and a 17,000 liter, steel reservoir in a small heated shed on the lakeshore. The operator hauled water to buildings in a 2000 liter truck. People without water tanks in their houses carried water from the lake in buckets. Many of them boated six kilometers off shore to obtain clear, soft lake water.

In late winter the Hay River stops flowing. Consequently the pump drew clear, lake water.

I returned to the community in 1956. At the suggestion of local people, I recommended extending the water intake pipeline. The open end should be beyond the swath of colored water from the Slave, Sandy, Buffalo and Hay Rivers.

DI&NA engineers asked Don Stanley, the community's consulting engineer, for his suggestions. He proposed driving a sand point—a pointed steel pipe with fine slots in the walls—into the sand on the north shore of the Island. Water from the Lake might seep through the sand and fine slots into the well.

Figure 161. *Yellow water of Victoria Falls on Hay River, near the town. Photo by Jack Grainge.*

However, ground water from the Island, not lake water, seeped into the well. The water contained high concentrations of dissolved iron and manganese. Stanley installed a treatment system on trial. Its operation showed that the treatment would have been too costly. If the well had been successful, it would have saved the high cost of constructing a long pipeline into Great Slave Lake.

In 1964 Beacher Linton Construction Co. laid a 6.4 km-long, 250 mm, steel pipe north into the lake. The innovative John Pope was the foreman. Previously he had shown his genius when he salvaged a tractor from the bottom of Kirkland Lake, near his home town. Being the only bidder, he bought it cheap from the insurance company. From a boat he hooked the tractor. Taking advantage of the buoyancy of the water, he raised it high enough out of the muddy lake bed to drag it near the shore. He hauled it out with an on-shore tractor.

Years later an NTCL barge had been blown ashore on Great Slave Lake. A Company crew had tried to tow it away but failed. Operating a water pump on the deck of the barge, Pope hosed away much of the ground from under the barge. Within a few days he towed the barge away.

At Hay River during the early, warm spring weather, Pope welded pipe sections together on the ice, to form a 6.4 km long, cement-lined, 250 mm pipeline. Then he rigged two circular saws to run off the power takeoff of a bulldozer, and cut a channel through the ice. He capped off both ends of the air-filled pipe and rolled it into the water, where it floated. Then he removed caps, one on the lake end and one on the shore end. The pipeline sank snake-like to the lake bottom. Without the advantage of an engineering degree, Pope was a super engineer.

The pipeline was long enough to obtain clear water throughout the year except for three weeks in spring when the river was in flood and the lake still ice covered. We had erred in basing our decision regarding the distance to the clear water on summer observations. We should have taken samples of the water from below the ice during the late spring.

In the pump house the water was chlorinated, but not clarified. The high cost of clarification of colored water for a three-week period in spring was not warranted. After I had retired, an eight kilometer-long pipeline was constructed in a different location and in another direction. Still the water before the ice cleared is colored.

* * * *

To go back in time, in 1951, many people had septic tanks and seepage pits. I found that the effluents from many seepage pits were surfacing. The hotel septic tank served no purpose; the effluent rose to the surface, creating a puddle outside the rear door.

I started to make a survey of all the septic systems, noting the sizes, depths and numbers of people served by each one. I thought there might be a solution, but soon decided that septic systems in Hay River were hopeless. The only feasible solution was a sewer system.

During an inspection in 1962, I discussed with local people the probability and consequences of floods over the Island and possible solutions. Everyone I spoke to agreed the land surface should be raised 1.5 to 2.5 m. There was enough sand off the north side of the Island that could be dredged hydraulically. During spring floods of the Hay River, that sand had been gouged from its banks upstream and deposited at its mouth. A second delta island was in the making, Vale Island being the first. In my report I stated that a large, floating, hydraulic dredge would be required.

I gave two reasons for suggesting the raising of the ground level of Vale Island. The first was to avoid the possibility of future flooding. The second was to provide a stable base for the construction of subsurface, piped water mains and gravity sewers and paved roads and sidewalks. I should have added that with a cover of fertile soil, the dusty town could have become green, attractive and dust free.

The cost of moving the houses during the sand-fill operation would not have been great. All of them were small, wood frame buildings without basements. Most were shacks. The hotel was ready for demolition. As a less desirable alternative to raising the ground level of the settlement, I suggested that a new town should be constructed on the nearby mainland.

By coincidence, in late May the following year the built-up section of the island became flooded to a maximum depth of 2.3 m. The local RCMP Sergeant performed admirably by waking people and warning them to get to high ground at the airport. They were evacuated to Fort Smith.

Figure 162. *Stores on the main street shifted around during the flood. Photo by Jack Grainge.*

To keep his basement free of seepage water, the builder of one house had enveloped the wood-frame basement with polythene film. The flood in the area was only a meter or so deep, but with a watertight basement the house popped up. The owner had an inland house boat, tilted but sound.

At the time I was in Taloyoak. When I reached Hay River a few days later, the flood was over. An engineer from my office had spread chloride of lime wherever sewage from septic tanks and seepage pits had surfaced.

A combination of events had caused the flood. Spring had arrived late in northeast Alberta and nearby NWT, the headwater regions of the Hay River. When snow-thawing weather finally arrived, it was unusually warm, so the thaw was fast. A flood of water and ice chunks bashed into the meter-thick, solid ice covering Great Slave Lake. Water could not escape down the back channel because it had been dammed by the causeway constructed in 1951. A bank of ice chunks jammed the river mouth, diverting the flood water over the island. Winnipeg has a similar problem with the north-flowing Red River.

DI&NA hired Stanley, Grimble and Roblin Ltd. to study the flooding problem. Stanley proposed three remedies. First, replace the causeway across the westchannel with a bridge. Second, block off the small channels between the small islands on the south shore of Vale Island. He thought that this would channel most of the water and the ice floes down East Channel, the main channel. Third, move the residential development to the nearby mainland. The engineer opened the causeway and replaced it with a bridge. The planner designed a new subdivision on the mainland

However subsequent floods of Vale Island were as bad as, and sometimes worse than, the one in 1962. The planners had failed to note my study of the causes of past floods and the likelihood of future high floods. If they had done so they likely would have agreed that raising the level of the island would have been worthwhile. Don Stewart, who later became mayor, was disgusted with their decision.

* * * *

One day Don Stanley phoned me to say that the town on the mainland would be restricted to a narrow strip between the proposed railway and the river bank. The result would be a long, narrow town that would be costly to service with water and sewer systems. The business section would occupy the middle section and the residential areas would be limited to the two ends of the town. He thought a wider town, with the schools, churches, stores and offices surrounded by houses, would be better. It would be less costly to service and more accessible for everyone.

Diverting the railway to double the width of the Town would have been simple. It would have added little to the length of the track. The residential district served by subsurface, piped water and sewer services would have been doubled. The extended town running along the high river bank upstream would have been shortened. The systems of water mains and sewers would have been much simpler and far less costly to construct, maintain and prevent from freezing.

I wrote a supporting letter to my head office. I presume they forwarded a copy to DI&NA. Stanley traveled to Ottawa to press the importance of accepting his suggestion, but the planners did not take his advice. Being far away from the decision-making committee in Ottawa, I had no opportunity to do more.

Years later, the Town's mayor and council complained about the decision to not move the railway. The editor of the local newspaper accused me of failing to support Stanley's recommendation. I forwarded the mayor a copy of my letter to my head office.

That action took the heat off me, but the Town will suffer forever. Many people who could have built houses in the fattened town built them upstream, from which they needed to drive to work, school and stores. I told local authorities that they should not allow development beyond the sewers.

Householders beyond the sewer system would construct septic tank systems, many of which would leak, often intentionally, directly to the River.

* * * *

In 1964 Beacher Linton Construction Co. constructed water and sewer systems for the mainland subdivision. To prevent freezing, they set the water mains with three meter minimum soil cover. Permafrost did not present a problem either during construction or since. Apparently any permafrost along the pipeline routes melted during the construction period. The fiberglass pipelines are strong and have had few breakages.

A 1,350,000 liter, steel, above-ground, cylindrical water reservoir and distribution pump station was built on the mainland. The reservoir was covered with a 9 mm coating of Insulmastic, a tar-based, waterproof compound which provided some insulation. One winter the tank overfilled. The water ran down the outside and froze. The vent froze and when water drained out, the tank imploded (sides caved in). The following summer the operator pumped water into the tank, compressing the air above it and thereby blowing the tank back into shape.

Figure 194. *Water overflowed reservoir in winter. Photo by Jack Grainge.*

* * * *

During construction of the new town, families lived in a group of about twenty trailers located on the mainland. The contractor bulldozed an excavation for a temporary sewage lagoon, with no outlet other than seepage. During midsummer, I was surprised to discover upon testing that the water was more than 100% supersaturated with oxygen. I concluded that the algae, with twenty-four hours a day of sunlight and twilight, was responsible for this extremely high concentration of dissolved oxygen in the water. Under such circumstances, treatment of the sewage was much faster than is usual in sewage lagoons.

Stanley and I discussed what would be the best sewage treatment system for the town. We decided there should be two small lagoons to remove the solids. The effluent would flow through wetland to Great Slave Lake, downstream of the town. The ponds should be connected so that they could be operated in either parallel or series.

The system operated well except that the Town's sewage truck could not be driven up to the lagoons. Consequently the drivers dumped untreated sewage at the lagoon outlet. The wetland satisfactorily treated both the untreated trucked sewage and the effluent from the pond.

In 1970, Dr. Jim Riddick, a senior engineer in DI&ND, allocated $100,000 for me to conduct research. My projects had to be beneficial to the companies constructing the oil pipeline from Norman Wells to Alberta. I accepted and promptly hired Dr. Richard Hartland-Rowe, Professor of Biology, University of Calgary, to conduct a study of the efficiency of this wetland in treating sewage.

Hartland-Rowe, hired P. Wright, a graduate engineer working on his masters' degree, to conduct a thirteen-month period of field tests. Dr. Roger Edwards a biologist and John Shaw, an engineer, both in my office, and I followed the field work and the chemical and bacteriological testing. The outflow from the wetland receiving the sewage was no different from that below nearby wetland receiving no sewage. Fantastic.

I told Hartland-Rowe that this was the first study of its kind. It was possibly the most outstanding single development in the science of sewage treatment. It warranted publication in a technical journal, preferably the prestigious *Nature*.

I was transferred to other work. The engineer who took over from me did not realize what a break-through in sewage treatment Hartland-Rowe had made. They did not publish his report in any journal. Following his final report in a 1974 Environment Canada publication, a rash of technical papers on the same subject by other scientists appeared in technical journals. Of course other engineers had been doing the same as we, namely, discharging sewage effluent into wetland. Obviously they followed our cue, and made similar studies of their wetland treatment of sewage, and published their findings in recognized journals. We had goofed.

* * * *

In 1976, Hay River Council decided that they needed to expand the town across the railway tracks to cover the area where the lagoons were located. Therefore Stanley's engineers designed new sewage lagoons farther away from the town. Construction could not begin until the existing sewage lagoons had been drained, so the contractor cut a ditch through the embankment. The contents of the lagoon flowed into Great Slave Lake.

The local environmental health officer (sanitation inspector), a DNH&W employee, complained about this action of the contractor to the mayor and to the editor of the local newspaper. On my return to Edmonton from a trip north, I happened to stop at Hay River. The DNH&W environmental health officer asked me what should be done. I replied that the contractor could not construct the new lagoons until the ground had drained dry. Therefore the flow into the lake had to continue. If people were living along the lake shore between there and the outlet to the Mackenzie River, he, the sanitation officer, should warn them that the water was contaminated.

Figure 164. *Hay River Delta. Photo by Jack Grainge.*

The environmental health officer was complaining that something should be done; other officials were taking samples of sewage at several points along the discharge ditch. I learned later that the local newspaper editor decided that I was the only one doing nothing, so I was his bad boy. I was the wronged person at the wrong place at the wrong time.

* * * *

In approximately 1969, DOT engineers asked me for suggestions regarding what to do about the disposal of sewage from the airport. I gave them suggestions for a sewage lagoon. They constructed a half-hectare-size sewage lagoon, one hundred meters northwest of the airport building. The effluent flowed to a 2.5 km long lake, a former channel. The lake snaked through the uninhabited middle of the island to the mouth of West Channel. I considered that lake to be a wetland providing superior prolonged treatment to the small overflow from the lagoon.

In approximately 1974, a government official required them to construct an aeration plant to treat the sewage prior to its discharge to the lagoon. I phoned the Regional Director, DOT, telling him that the sewage was already receiving superlative treatment. The treatment plant was built, but it never operated properly. After I retired I heard that the treatment plant was abandoned.

* * * *

In 1965, Penticostal Assemblies constructed a new large, two-story, wood-frame, H.H. Williams Memorial Hospital in the new town. For a foundation the ground was covered with a meter depth of gravel, with an extra third of a meter of gravel below concrete beams supporting exterior bearing walls. The soils engineer drove test stakes four to five meters into the ground where they struck a hard layer of either sand or gravel. However, below that hard layer was four meters of permafrost containing ice layers. The contractor drove timber piles supporting interior bearing walls to that first hard layer. A year after construction, heat from the hospital melted the permafrost, and the building settled unevenly.

Franki Canada Ltd. was hired to underpin the building with mega piles. During the winter of 1968-69, George Stefanick, an engineering classmate of mine, and his assistant did the work. They hand-dug a working space under the building. One after another they jacked down the original piles, welded eighty-centimeter sections of steel pipe to it and jacked them down. They continued welding additional sections together and jacking them down until they reached adequate refusal depths at 11.3 to 12.7 m.

* * * *

The completion of the road from Hay River to an Alberta highway in 1947, made possible commercial fishing on Great Slave Lake. Five companies built plants for gutting and packing the fish in boxes for shipment to northeast United States. Five plants were in Hay River and one at Gros Cap, an island in the lake. Local fishermen in small boats netted fish and sold them to the plants. Menzies Co. had a couple of barges on the lake with a staff gutting and packing the fish. When the fish passed through Edmonton on their way east, fisheries inspectors in Edmonton examined the fish for quality and worms. United States federal inspectors at the Canada-USA border also checked them for similar conditions.

In July 1966, there was an outbreak of Salmonella java in New York, New Jersey, Connecticut and Pennsylvania states. The first outbreak of the disease was at a barmitzvah in New Jersey. The source of the disease was smoked fish from Hay River, which had been processed in United States plants.

My chief in Ottawa asked me to meet three epidemiologists flying to Edmonton. Dr. Eugene Gangarosa, Center for Disease Control, Atlanta, Dr. Alan Bisno, New Jersey Department of Health, and Dr. John Murphy, epidemiologist, from DNH&W, Ottawa.

The three epidemiologists and John Shaw, an engineer in our office with experience in epidemiology, flew to Hay River. They found the water pumped from the lake in front of the plants to be contaminated. They found none of the staff to be a carrier of the disease.

In his report Dr. Gangarosa stated that none of the rectal swabs taken from the women gutting the fish and the swabs of the cutting tables showed signs of any type of salmonella bacteria.

Figure 165. *Fish cleaning and packing plant at Gros Cap. Photo by Jack Grainge.*

His report contained information about some workers in the Chicago plant being carriers of the disease. He also criticized some unsanitary practices in the Chicago plant, such as failing to smoke the fish adequately.

A few days later Frank Gillis, the local Environmental Health Officer, John Shaw and I flew to Gros Cap to examine that plant. While waiting outside for our plane to take us back to Hay River, we overheard three little Dene boys talking. They were wondering whether John Shaw, with his full beard and fairly long hair, was a man or a monkey.

In the late 1960s a Fisheries officer from Ottawa invited me to a meeting with representatives of the plants in Hay River and Gros Cap. A portly representative of a processing plant in Chicago also attended. He criticized the Department of Fisheries officials, sparing no one's feelings. During coffee break, the manager of one of the plants in Hay River said to me, "If one of us were to ship fish to a processing plant other than that fellow's, he would fly to Hay River to loudly bawl out the manager. The fellow would shout that the next load of fish shipped from that manager's plant would be rejected by USA inspectors at the border." The fellow's predictions were always true. My informant inferred Mafia connections.

The next year at a meeting in Hay River, a Department of Fisheries officer reported new, strict sanitation standards that must be met by the plants in Hay River. The operators asked me to attend a meeting regarding what they could do. I explained that complying with the standards would require costly equipment. The Department built a plant on the mainland in Hay River and the privately owned plants closed. The barge owners sold their catches to the new plant. That spelled the end of the first private industries in Hay River.

References:
Camsell, C. 1954. *Son of the North*. New York, N.Y.: D. McKay Co.
Dusel, F. 1999. Personal communication.
Edwards, B. 1997. Personal communication.
Piard, J. 1997. Personal communication.
Shaw, J. 1997. Personal communication.

PINE POINT

In May 1952, the Department of Indian and Northern Affairs (DI&NA) administrator of western NWT, based in Fort Smith, planned a meeting at the site of the future Pine Point community. Consolidated Mining and Refining Ltd. (COMINCO), owners of Pine Point Mines Ltd., was planning a new, company-owned town. Seven of us, including Aubrey Perry of our Vancouver office and me, met with the regional director in Fort Smith and then flew in a single-engine Norseman to Pine Point.

Clair White, the superintendent of Con Mine in Yellowknife as well as this one, met us. We rode in the back of a light delivery truck the five kilometers to the proposed townsite. It was a grassy plain, an esker of gravel, sand and silt underlain by clay. It was surrounded by forest bluffs.

There was adequate space for a large town. The clay subsoil was suitable for laying deep water mains and sewers. Although permafrost may have existed, the soil was not subject to frost heaving. A 32 m deep well produced an adequate water supply. The water was hard but satisfactory. Unfortunately neither Perry nor I asked questions about the geology of the water aquifer. There was adequate space for a lagoon. Perry and I reported that the area was a good site for a settlement.

Two years after our first trip, the new superintendent of Con and Pine Point Mines flew a few other engineers and me to Pine Point. While we were waiting for a return flight, I mentioned to the representative of Hays Development Co. a word about their company buying the farm of Fred Percival, Earl of Egmont, bordering Calgary. He replied, "Yes. We met him in his blacksmith shop where he was working. We agreed on a purchase price of a million and a half dollars (a fortune in 1952). I added that we would make three payments of half a million dollars spaced over a year." Percival drawled quietly, "No. Then I'd have to get dressed in a suit three times to go downtown to see a lawyer. Make it one payment or the deal's off."

A few years later, Jim Donaldson, COMINCOs buildings design engineer, asked me to help him plan the sewage treatment system. He knew how to design the water system and the sewers. They were planning a community with a population of one thousand.

I recommended a sewage lagoon conforming with Alberta provincial regulations, a single cell, sized according to a scale of one hectare per 250 people. To compensate for thicker ice on an NWT lagoon and consequently less volume, I suggested a depth of 1.8 m, 0.3 m deeper than Alberta's standard. Donaldson designed a lagoon of somewhat more than four hectares. The effluent overflowed to wetlands, which extends twenty kilometers to Great Slave Lake.

*   *   *   *

The lead and zinc were mined by an open-cut operation. Thus the mine consisted of a huge, deep excavation. The ore was in a porous, fractured limestone formation below glacial till. Ground water seeped through the porous limestone into the excavation. To keep the excavation dry, pumps drew water from dewatering wells around the perimeter of the excavation. As the mine excavation went deeper, they drilled the dewatering wells even deeper.

Before the dewatering, the aquifer tapped by the town's water well produced surface water. That fresh water floated on the deep salt water of a preglacial sea. It had been there since the limestone coral-reef stratum formed The dewatering wells around the mine excavation drained the water from the aquifer tapped by the town's wells. Therefore the town foreman drilled successively deeper wells, eventually reaching a depth of 245 meters, much lower than sea level. The water obtained was a mixture of waters from the ancient sea and surface seepage water. It was salty, much of it sulfates. The sulfates in the

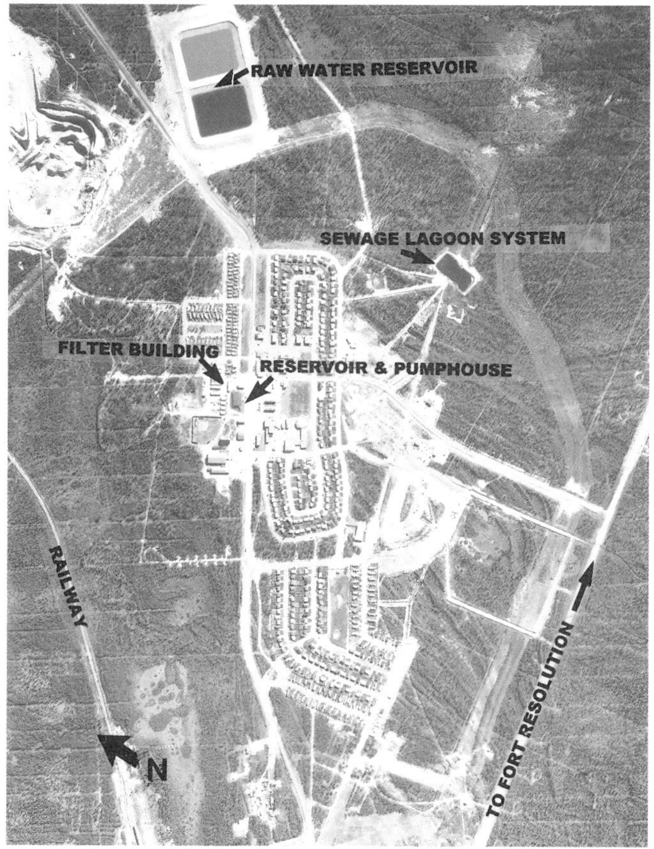

Figure 166. *Aerial photo of Pine Point. Department of Energy, Mines, and Resources Canada, 1964.*

wastewater combined with organic matter in the sewage lagoon to produce mercaptans, the stinkiest of substances.

Four hundred meters of forest separated the town from the sewage lagoon. I had thought that would be a satisfactory odor barrier. However the construction of the sewer to the lagoon resulted in a corridor through the trees. During the winter while ice covers lagoons, anaerobic (without oxygen) decomposition occurs in them. The unoxidized sewage contained mercaptans, Phew! When the ice on the Pine Point lagoon melted and winds blew the stench toward the town, the residents cursed the designer of the sewage lagoon. When I went to Pine Point, I did not mention my part in the design.

The sulfates (Epson's salts) and malodorous hydrogen sulfide, as well as lead in the drinking water rendered it unfit for any household purpose. The Mine installed a reverse osmosis demineralizer. Osmosis is the process of water containing low concentrations of salts flowing into a dense root

Figure 167. *Pine Point lagoon. Photo by Jack Grainge.*

Figure 168. *Pine Point Mine. Photo by Jack Grainge.*

through its slightly porous skin. In the case of reverse osmosis, water under high pressure is pressed through a slightly porous membrane. The membrane holds back some minerals including the sulfates and lead. The rest of the water flows to waste.

I had not visited Pine Point during the mid 1960s when the water system was beginning to give problems. I suppose that for a while the demineralizer produced satisfactory water. In the mid 1970s I went there regarding contamination of the water. After dealing with that problem, the operator and I discussed the demineralizing process. The well was 345 m deep. The system could not reduce the concentration of salts sufficiently. Also the cost of the operation was too high. I suggested that they consider piping water from Great Slave Lake.

The Company asked Stanley and Associates Engineering for an opinion. Don Stanley, Bob Dawson, then in Stanley's firm, and I went to the community where we met with government officials. The question was whether to increase the size of the expensive demineralizer or to construct a twenty kilometer pipeline to Great Slave Lake. Of course the pipeline was cheaper and infinitely better.

I had no occasion to go to Pine Point again. The quality of water was much better. I heard that the malodorous conditions of the sewage lagoon improved. I felt better about that.

Since my retirement, the mine closed and the people have gone. The town and municipal problems are now only memories.

Reference:

Stelk, C. (Charley) R. 1998. Personal communication (Professor Emeritus, University of Alberta).

FORT RELIANCE

Fort Reliance is situated on a bay at the east end of Great Slave Lake. Lockhart River, which flows only in summer, enters the lake nearby. When I went there in the mid 1960s it consisted of a DOT weather station and multiple residence and a fishing lodge a short distance away.

The DOT had built a staff residence with running water systems. Their engineers in the Edmonton office asked me for suggestions regarding what to do with the effluent from their septic tank. The fishing lodge had no running-water sewage system.

Figure 169. *Aerial photo of Great Slave Lake area, situating Fort Reliance, fishing camps, and DOT weather station. Department of Energy, Mines, and Resources, 1970.*

Both the weather station and the lodge pumped water from the bay. Due to the circulatory nature of wind-caused currents in the bay, septic tank effluent should not flow into the bay. The watertight gravel and clay ground had neither filtration nor absorption capacity.

Figure 170. *Dog team at the DOT single men's residence, March 1969. Photo by Jack Grainge.*

In looking for a site for a sewage lagoon, we found a large basin within the Precambrian rock about forty meters from the DOT buildings. It had a small amount of grass and lichens at the bottom, probably fed by rain. The wet bottom showed that the basin was watertight.

I suggested that they install a tank to receive the effluent from their septic tank, an effluent pump and an aboveground, insulated pipeline leading to the basin. The pipeline should be laid to drain after each pumping. The pipe boxing should contain an electric heating cable. Before pumping, the electricity should be turned on to warm the pipe and not turned off until after the pipe had drained.

We expected that the basin was large enough that all the effluent would evaporate. Mike Greenwood, the meteorologist in charge of the station when it closed during the 1980s, reports that the sewage disposal system had operated well.

Now the buildings are for sale. I understand that it could be made into a resort for nature buffs and fishermen.

LUTSEL K'E (SNOWDRIFT)

Lutsel k'e is situated on a peninsula jutting into Christie Bay on the south side of the east end of Great Slave Lake. The land is low, bordered on the east by forested, high rocky ground. In 1925, the HBC established a trading post on a small but pleasant meadow at the north-west end of the peninsula. In 1954, Dene, who had lived nearby for many years, began moving to Lutsel k'e. During the 1960s, the Department of National Health and Welfare (DNH&W) built a nursing station on another meadow on the south side of the peninsula. At the time, the boggy land between the two centers was unoccupied. Access between the two centers was over bare precambrian rocks near the shore.

In approximately 1957, with a public health nurse, I visited the community on a charter flight. There were only the HBC employees and a few Dene living there. They bucketed water from the lake and discharged washwater nearby. They discharged garbage and emptied honey buckets some distance inland.

Figure 171. *Hudson Bay Company (HBC) outpost (1963) at Snowdrift. Photo by Jack Grainge.*

During the mid 1960s, Dr. Otto Rath, the Regional Director, Medical Services Directorate, DNH&W, in Edmonton, asked me to go there to make suggestions regarding the nursing station's water and wastes. Once again I accompanied a nurse. She was returning a baby from the hospital in Yellowknife to the family. A government administrator sat in the front with the pilot. The nurse told me that on the previous flight the same two had been in the front while she was in the back holding a baby. For some reason that she did not understand the plane on floats, after landing on the water, began to sink. The two in the front opened the doors on their respective sides and escaped, leaving her in the back holding the baby. Fortunately the plane settled on the bottom without being completely submerged. She was able to hold the baby's head and her own above water until the pilot helped the baby and her to get out. She felt that the men had jeopardized the baby's and her safety. Possibly they momentarily forgot about her.

At the time there were only a few Dene and the HBC and DNH&W employees living there. As noted on the previous visit, everybody carried water from the lake. They discharged liquid wastes near their houses, and carried garbage and toilet wastes inland.

Figure 172. *HBC Manager and candled ice. Photo by Jack Grainge.*

We landed at the HBC dock and I walked along the rocky shore to the nursing station. I cannot understand why that location for the nursing station was chosen.

Figure 173. *Fishing lodge on shore, 1965. Photo by Jack Grainge.*

I cannot remember but I probably suggested that they pump water from some distance offshore in the lake and discharge wastewater near the station. I am sure that I saw no alternative to the use of honey buckets. I remember that I did not suggest disposal of sewage to the lake. The circulatory currents in the lake would flow unpredictably in different directions depending on the winds. The only reliable source of safe water would be well offshore, requiring use of a boat. This would entail much handling and consequent, inadvertant contamination.

I do remember that it was raining when I returned to the HBC dock. The rocks along the way were slippery and I slipped and fell, breaking a fibula bone. It was a painful walk back and the injury took a few weeks to heal.

In 1976, Northern Health Services had an extended-aeration, sewage treatment plant installed to serve the nursing station. I had previously advised our department supervisors that such systems were difficult to operate and served no useful purpose. If operated well, they reduce the oxygen-demand of the sewage, but they do not reduce the sewage bacteria significantly. Chlorinators in a remote system such as this, are not reliable. The lake had an over-abundance of dissolved oxygen, but sewage contamination of the water would continue to be the problem.

Now the community has grown. Roads have been built, and sewage and water are hauled. I should have realized that, like other northern communities, this one would grow. When I was first there, the community was only a small HBC fur trading post and two or three houses. At that time I should have recommended looking for a better site for the community.

References:
Shaw, J. 1968. Personal communication.

FORT RESOLUTION

Fort Resolution is situated on a rock outcrop surrounded by a grassy plain on the shore of Great Slave Lake. It is a kilometer or more south of the mouth of the Slave River. Much of the surrounding shore is swampy. The soil is loam, underlain by fine sand and silt.

In 1876 Cuthbert Grant and Laurent LeRoux built a North West Co. trading post on the Slave River shore nearby. A few years later they moved their post to Moose Deer Island, and the HBC built a trading post at the present site. The two companies merged in 1821 and used the present site.

On their arduous return from Coppermine in 1821, Sir John Franklin and Sir George Back wintered at the fort. By hunting throughout the famine-plagued winter, the indomitable Chief Akaitcho saved the lives of both the explorers and his own Dene.

In 1852 Father Faraud established a mission in the settlement. About forty years later the mission built a boarding school, and in 1938-39 an eighty-bed tuberculosis hospital. The RCMP established a post in 1913. In 1971 the government extended the gravel road from Pine Point to Fort Resolution.

I first went there in 1966. Pat Neven, a cheerful, Irish engineer in the Department of Indian and Northern Affairs (DI&NA) office in Fort Smith, was a great help to me. He and I were always good friends. Once he happened to be in Edmonton while Indians from a reserve near Edmonton staged an orderly demonstration in front of the federal building. He said to me, "Its too bad they don't have any Irish blood. If they did we could be watching a good fight."

Figure 174. *Early aerial view of Fort Resolution, 1956. Photo by Jack Grainge.*

The population of Fort Resolution in 1966 totaled 750, including 450 Métis and 250 Dene. The center of the community contained a two-man RCMP detachment, nursing station, school, diesel-electric power plant,

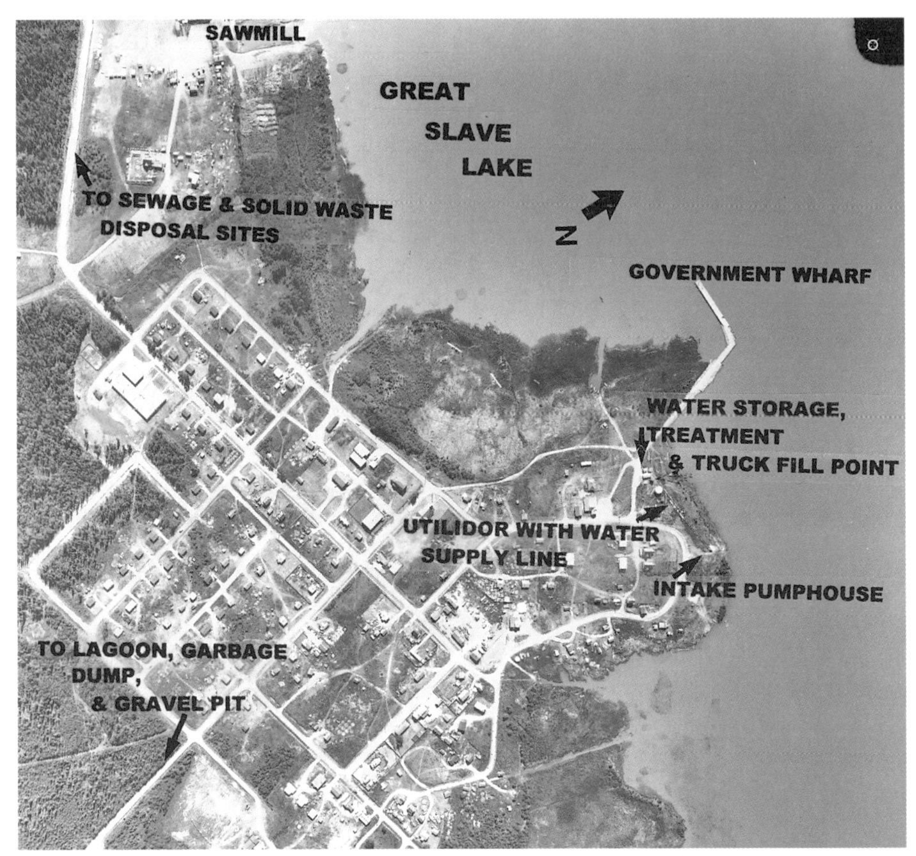

Figure 175. *Aerial photo of Fort Resolution. Department of Energy, Mines, and Resources Canada 1979.*

radio communications station and several houses. To the northwest was a DOT airport, and family houses for two airport personnel. To the southwest was a beautiful Catholic mission and their almost empty, eighty-bed hospital. Before the construction of adequate facilities in the Camsell Hospital in Edmonton, the Fort Resolution hospital had been full.

A cooperative-type saw mill, a half kilometer north, employed twenty or so local people.

* * * *

The only source of drinkable water was Great Slave Lake. It was Slave River water, somewhat hard and in summer turbid. Several people had drilled shallow wells only to obtain extremely hard, highly colored, saline water. The people hauled water for themselves but gave their cows well water. I took samples for analyses. When I saw the high content of Epsom salts in the well water, I understood why the cow corral was awash with muck.

Most people carried their own water, but the government contracted a man to deliver water from the lake using a 4500 liter tank truck. He added chlorine solution each time he filled the tank. Most people kept water in abandoned oil barrels in their kitchens. Within two days the water near the surface would clarify. They then dipped it from the barrel. A few people boated five kilometers offshore to obtain clear, lake water. From October to May, the Slave River becomes frozen over and the water clarifies.

Figure 176. *The teacher used a filter to obtain clear water. Photo by Jack Grainge.*

I suggested that a good, cheap way to obtain clear water for a public system would be to have two 130,000 liter tanks. Use settled water from one while turbidity was settling out of the water in the other. If only one tank were constructed, clear water could be obtained by using a floating, skimming type of outlet. Since the water was turbid only in summer, the tanks would not need to be protected from freezing.

* * * *

Most of the houses with running water systems used septic tank systems. The wastes water seeped away through the sandy subsoil. The houses near the dock were on rock, the primary reason for the HBC choosing that location. It was a good place to build a dock and a dry site for buildings, but a poor place for seepage of sewage.

Sewage from the rectory flowed into a lagoon near the shore, from which it overflowed to the lake. Dense water weeds extended a half kilometer off shore. The weeds prevented short-circuiting before there was some digestion of the sewage. Nevertheless, according to our bacteriological tests, the water

Figure 177. *Hauling water from the dock. Photo by Jack Grainge.*

downstream of this point was contaminated, possibly by ducks. I suggested that a sewage lagoon be constructed.

Most of the people used poorly constructed privies, and discarded waste water beside their houses. I was there in late September, after the fly season. Flies must have been a problem during the summer. I suggested that they build privies according to a plan that we had prepared. We received no requests for copies. I should have signed the plans as "The Specialist." The article, The Specialist, contained hilarious instructions for building a privy. The author even made a special box to hold last years Eaton's catalogue. He estimated that by Christmas the people would be into the hardware section.

The contractor hauled garbage to a dump site almost a kilometer north, off the road to the saw mill.

A month after my examination of the community, Norm Lawrence of AESL, made an on-site study. He made

Figure 178. *Church and rectory sewage lagoon. Photo by Jack Grainge.*

Figure 179. *Daughter of Dr. Greendige and vessels at dock. Photo by Jack Grainge.*

a series of suggestions for community development to be introduced as the community grew. However it did not grow much. Lawrence agreed with me. For the community of 750 people, settling the turbidity during the summer in community tanks was a reasonable solution.

Five years later I returned to the settlement. I was surprised that more wells had been drilled. Since my previous visit a new large school had been built. DI&NA was building many water and sewage systems. Therefore I restated one of Lawrence's recommendations. I suggested that they build water mains and sewers to serve the school and other buildings with running water.

References:

Grainge, J.W. 1966. *Report on Sanitation, Fort Resolution, NWT.* Edmonton: Public Health; Engineering Division, Dept. of National Health and Welfare.

Grainge, J.W. 1971. *Report on Sanitation, Fort Resolution, NWT.* Edmonton: Public Health; Engineering Division, Dept. of National Health and Welfare.

Lawrence, N.A. 1966. *Fort Resolution, NWT, Engineering Report.* Yellowknife: Associated Engineering Services Ltd.

FORT SMITH

Fort Smith is situated on a sand plateau, thirty-eight meters above the Slave River, downstream from the Rapids of the Drowned, the last of four rapids on the river. It is a good fishing, trapping and hunting area, so probably Dene lived in the region from prehistoric times. Early traders came to the vicinity as explorers: Samuel Hearne in 1772, Peter Pond in 1780, Alexander Mackenzie in 1788 and Sir John Franklin in 1820. In 1852, Father Vital Grandin (later bishop) established St. Isadore Mission at Salt River, 32 km downstream from Fort Smith. In 1876, Father Zephyrin Gascon moved that mission to Fort Smith. The missionaries taught the Dene to read and write.

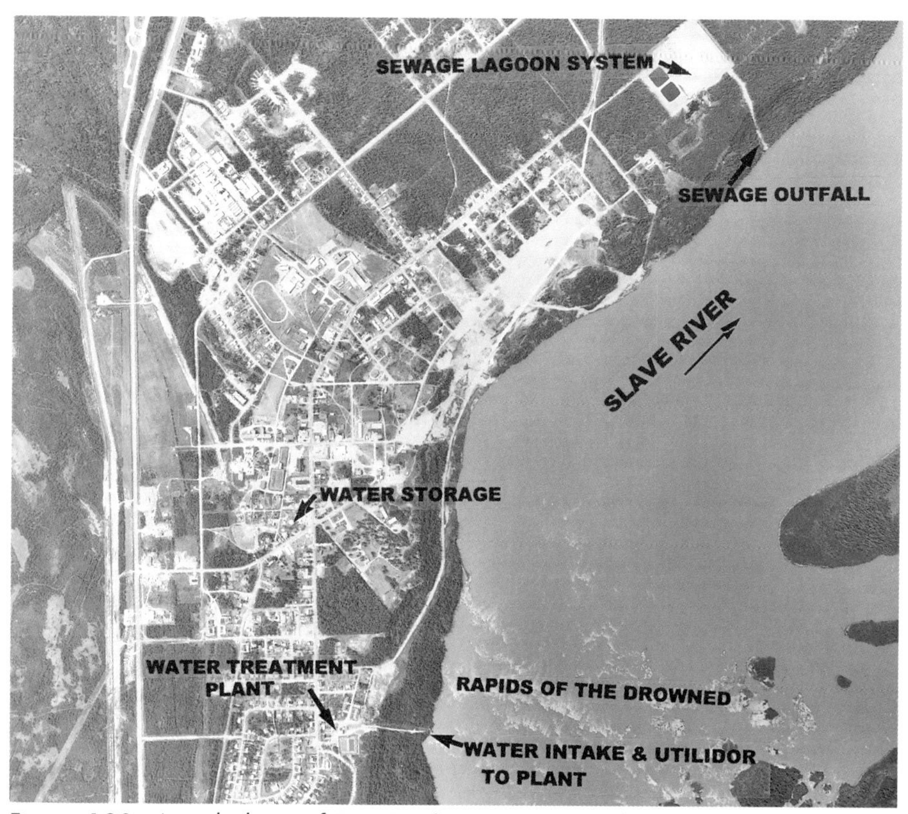

Figure 180. *Aerial photo of Fort Smith. Department of Energy, Mines, and Resources Canada, 1979.*

In 1874 the HBC established a trading post in a stockade at its present location. The Company named the post after Donald Smith, Chief Commissioner of the Company. Later Smith became the HBC governor and then the main financier for the construction of the C.P.R. railway from eastern Canada to Vancouver. He received the title, Lord Strathcona.

The Mission built the Grey Nuns Hospital in 1914, a one-room school in 1915, St. Isador's Church in 1923, a new St. Anne's Hospital in 1952 and St. Joseph's Cathedral in 1958.

In 1880, the HBC established a dock upstream, which they named Smith's Landing. In 1915, they renamed it Fort Fitzgerald, after RCMP Inspector Francis Fitzgerald. In 1911 he and three constables, had starved to death on a patrol from Fort McPherson to Dawson.

* * * *

In September 1950, I went to Fort Smith. I had just joined the federal Department of Public Works (DPW), and was asked to construct septic tank systems for federal buildings. On the road in from the airport, I asked the taxi driver why the road was sand with no gravel. He said, "You think this road is bad, wait 'till you see the one south to Fort Fitzgerald."

The 26 km road between Fort Fitzgerald, Alberta and Fort Smith had always been a bottleneck. In early days men on shore usually tracked (pulled) York boats both up and down the four rapids on the Slave River between those communities. Then the HBC cleared the trees to make a 26 km portage. They hauled freight over the muskeg with oxen, but often in summer horse flies and mosquitoes were too much for the animals.

Figure 181. *John Piard and daughters of Steve Yanish holding mounted deer heads. Photo by Jack Grainge.*

During the late 1920s Mickey and Pat Ryan, working on a freight boat arrived in Fort Fitzgerald. Mickey had ended his boxing career in Edmonton where he broke a hand in a fight. The Ryans began hauling freight across the portage. They used horses wearing nosebags to keep away the horseflies and mosquitos. To improve efficiency, the Ryans established a camp on a small prairie bordering the river, half way between Fort Fitzgerald and Fort Smith. Thus a driver would begin in the morning hauling freight to either port and be back with a different load in the evening. They aptly named their settlement Half Way.

The Ryans built a corduroy road by laying tree trunks across it. As some trunks sank out of sight in the mud, they piled more tree trunks on top until the road was somewhat firm. There being no gravel, they piled sand on top to make a road in which wagon wheels sank. With continuous rebuilding, hauling heavily loaded wagons, even during the wettest conditions, was difficult but possible. It thus became their private road, and competitors had to travel along the adjacent muddy road.

In early winter 1934, the Ryans portaged the 125-ton vessel Pelly Lake from Fort Fitzgerald to Fort Smith. They built a long heavy sleigh under the vessel, and hauled it with their two Linn tractors (tractors with tracks at the back and skis that are interchangeable with wheels at the front). A year later, they hauled the vessel Margaret A and four 100-ton barges across the portage. Later that year they hauled the Pelly Lake back to Ft. Fitzgerald.

My big, still-husky friend, Ab Coyne, was on the crew lowering those vessels down the thirty-meter-high river bank. They used seven stages of deadmen set vertically in holes to anchor one end of a block and tackle. The Lynn tractors controlled the other end.

While attending the University of Alberta, Coyne worked the summers of 1934 and 1935 for the Ryans. For working from 7 a.m. to 9:30 p.m., seven days per week, he received $75 per month, and three meals a day. For dinner he ate three heaping platefulls plus a pie and a half. Now in his nineties, he tells me he cannot understand what happened to his former, robust appetite.

Gilbert LaBine's Eldorado Gold Mines Ltd. bought Northern Waterways Co., a small, river transportation company owned by C. Murdoff and Cy Becker, an airplane pilot, and changed its name to Northern Transportation Co. Ltd. (NTCL). The company hired F. Corser and J. Doyle to build a road parallel to the Ryan road, which they completed in 1942.

At this time demands of the war played a vital role in the development of the area. Scientists in United States needed uranium, hitherto a waste product at Eldorado Mines, for building atomic bombs. Hon. C.D. Howe nationalized and expanded the mining company and its subsidiaries under President Bill Gilchrist, with Lionel Montpetit manager of NTCL and Alf Caywood manager of Eldorado Aviation Co.

American soldiers, using heavy equipment, joined both portage roads. They added tree trunks and tons of sand, and deepened the drainage ditches. Over the reconstructed portage they towed trailers heavily loaded with motor vessels, barges, oil pipe and construction materials. They gave the motor vessels and barges to NTCL to transport steel pipes and materials for the construction of an oil pipeline from Norman Wells to Whitehorse. They also built airports at Hay River, Fort Simpson, Norman Wells and one six kilometers north of Fort Smith.

The American army also drilled several new oil wells at Norman Wells, built a road and oil pipeline from there to Whitehorse, and an oil refinery at Whitehorse.

In an amazingly short seven months' time, the American Army constructed a 2473 km, all-weather, gravel road through the wilderness of bushland, swamps, mountain ranges and across rivers, from the railhead at Dawson Creek, B.C. to Fairbanks, Alaska. Parallel to this they constructed a chain of airports, including buildings and connecting roads and telephone systems.

After the war these highways, airports and oil wells have helped the oil, mining, forestry, and tourist industries, and in addition the development of the North. Imperial Oil Ltd. (IOL) transported the refinery to Edmonton, where it became the beginnings of a large, local oil industry. The people in the North salvaged much of the buildings and equipment, especially the barges.

The offices and warehouses of the directors and staffs for all of these operations were in Edmonton. In October 1942, the heaviest snowfall on record blanketed Edmonton, and it did not melt until spring. The City, at that time having no snow-clearing equipment, was paralyzed for three days. The American Army came to the rescue. It's crews made two snowplow passes along all main streets. Streetcars and cars could then get through. The merchants dug pathways from the cleared roads to their doorways. People tramped over the heavy snow on the sidewalks.

* * * *

In the early 1940s, HBC's Mackenzie River Transport Ltd bought the Ryan brothers' company. Ken Spencer was in charge of shipping from the railhead at Waterways to the arctic sea. He and his wife and two sons lived at Half Way. From earliest times the HBC wintered their vessels on the shore at Gravel Point, on the east shore of the river opposite Bell Rock, 11 km downstream from Fort Smith.

In 1947 Mackenzie Transport gave up its common carrier license, but kept its motor vessels, barges and ocean-going ships to haul freight for HBC trading posts. The company sold their freighting equipment and buildings in Half Way to NTCL. Spencer moved to Fort Smith. His house and his office were in an HBC quadrangle with their trading post, and the Mackenzie Hotel, all of them white, wood-frame, with bright red roofs.

Spencer was short of sober, reliable employees. Consequently he hired his fourteen-year-old son, Keith, a keen and bright non-drinker, as purser on a vessel. HBC regulations did not allow hiring relatives, so Keith was listed as Keith Robinson, his first and second names, age eighteen. Keith is now a criminology professor at the University of Alberta. Perhaps his early experience on the dark side of the law provided him with insight into the criminal mind.

The NTCL motor vessels were up to date with steel hulls, whereas Mackenzie River Transport had not replaced their old, wooden vessels and outdated equipment. In 1957 Mackenzie Transport discontinued all transportation operations and NTCL took over.

Spencer retired and moved to Edmonton where he had been spending his winters. He became one of the organizers of the Royal Glenora Club. He and I became acquaintances in Fort Smith and friends when I joined that club, of which he, upon retirement from the HBC, became secretary treasurer.

* * * *

In the early days at all three settlements, Fort Fitzgerald, Half Way and Fort Smith, water was carried, hauled, or piped from the River. Throughout the summer the river water was turbid, extremely so during floods. There are two floods, one in May and June when the snow in the forests melts, and a lower one in July when the snow in the mountains melts. People stored water in bottles or pails for a few days, to allow much of the turbidity to settle.

People at Fort Fitzgerald and Half Way used privies and threw waste water to the open ground. They discharged no wastes into the river. During the 1950s, I found the untreated water at the Fort Smith water plant to be free of bacterial pollution.

River vessels, with four-to-six-men crews, docked with prows pointing upstream. Thus the water intakes were ahead of the sewage outlets so the water in the boats was generally uncontaminated.

Each NTCL motor vessel had a filter to remove the turbidity particles from the water used for drinking and cooking. It was available from a third tap in the kitchen. The filter consisted of a manufactured, disc-shape, porous, diatomaceous stone. When the surface of the stone became plugged with silt, the cook turned a handle that rotated an internal knife scraping off the top surface of the disc together with the silt particles. The cook then turned on a water tap to wash away the scrapings. When the stone filter disc wore thin, he replaced it.

Unfortunately, in the early 1960s, the American company that produced these filters, discontinued supplying them. The NTCL had avoided installing drinking water reservoirs because the added weight might cause the vessels to become stranded on mud bars. Al Smith, the engineer in charge of NTCL water and sewer systems and I kept looking for a suitable substitute for the stone filters. We watched a demonstration of a centrifugal filter, but it did not remove turbidity from Saskatchewan River water. Better centrifugal filters are now available.

The Company installed drinking-water reservoirs, which they filled from Great Slave Lake, the Mackenzie River immediately downstream of Fort Norman or at towns where clear water was available.

* * * *

The population of Fort Smith was about four hundred. I stayed at the HBC's two-story, wood-frame Mackenzie Hotel. The hotel office, manager's suite, large dining room and kitchen occupied the main floor. Guest rooms and a common bathroom were on the second floor.

Louis Nordseik, a slim, pleasant and efficient manager, hired clean, friendly and hardworking employees. Most of them were Dene or Métis women. Mike Dempsey, formerly the first ranger and later superintendent of Wood Buffalo National Park, was the night watchman. One of his duties each night was to peel a pail full of potatoes. I used to help him and at the same time soak up his fascinating stories of early Fort Smith. He had married a Dene woman from Fort Chipewyan, and they had ten children. One, a pleasant, attractive young lady, worked in the hotel.

The hotel had been built to accommodate travelers who were awaiting connecting boats on opposite sides of the portage. The washrooms were primitive, but better than the northbound travelers would have in their houses in the settlements. When I arrived, people were traveling north by airplane, so the hotel was only for visitors to Fort Smith.

The hotel had running water and a septic tank system with sewage discharging half way down the river bank. There were flush toilets in the public washrooms. In about 1955 the HBC installed a private washroom for one room. That was for Governor General Vincent Massey who stayed there one night.

We did not get menus at meals. Everyone received the same delicious food. Nora Mandeville and Margaret, two local Dene women, were superior cooks. People never complained. With such good meals at one dollar each, how could they?

At lunch I met Bill Mann, a tall, bushy-browed, gray haired, gaunt, old Yorkshireman. He was the foreman of the crew building the federal houses and septic systems. His employees worked

Figure 182. *Pipeline up river bank. Trees topple down the bank when permafrost thaws. Photo by Jack Grainge.*

sixty hours per week for eight months of the year and collected unemployment insurance the rest of the year.

By the time I arrived the houses had been largely built. Abandoned houses at Abasand, near Fort McMurray, had been taken apart, the nails pulled, the boards of equal length strapped together and shipped to Fort Smith for reconstruction. We used the boards, studs and joists, but could not figure where to fit the short lengths. We burned them.

George Smith, the time keeper, gave me the plans for the septic tank and log-walled, seepage pit systems. All but one served individual houses. We buried them with a minimum soil cover of about 1.5 m. Because of the high frost protection required, the systems were deep. The soil in Fort Smith is mainly porous sand, which runs ten or more meters deep. The systems operated well.

* * * *

A crew, hired by the Department of Indian and Northern Affairs (DI&NA), was constructing a water treatment plant and water mains. The water intake was located at the end of the Rapids of the Drowned, the last of four rapids. It consisted of two electrically driven pumps in a wood-frame pump house near the edge of the river and a 200 mm, buried steel pipe extending into the river.

Figure 183. *First water intake below Rapids of the Drowned. Photo by Jack Grainge.*

Workmen told me that permafrost in the river bank soil had been a serious problem. First they stripped away the bushes and surface vegetation, and the permafrost melted. Then they manually dug a trench a meter and a half deep and laid the cast iron pipe. At the half way point along the pipeline, the ground slumped. The pipe broke so they inserted a short section.

Figure 184. *Replacing pipe-break with a zig-zag rubber pipe connected in an insulation box. Photo by Jack Grainge.*

The water treatment equipment consisted of two alum-mixing tanks, a sedimentation tank, a rapid sand filter, a chlorinator and a storage tank below the concrete floor.

A backhoe-and-bulldozer operator and two workmen laid the 200 mm, asbestos-cement, water mains. Apparently no one on the job had had experience in laying water mains. I had spent the previous three months as the engineer checking the construction of the water and sewer systems in St. Paul, Alberta. I realized that they were making mistakes. I told them that an experienced foreman on my previous job made sure that water mains lay straight and on smooth, compacted beds. Before bulldozing the back-fill earth, men hand-shoveled protective soil around the newly laid pipes. The workmen on this job said they had no such instructions, and continued to do the job their way.

I became acquainted with the foreman who told me he owned a successful construction company that was working on a contract in Quebec. He said he had a capable foreman on that job so he accepted the position of foreman on this job. I mentioned my misgivings about the water main's construction. He thought that the sand would make a stable bed for the pipe.

I also expressed my concern that neither of us used shoring along deep pipeline trenches. He said the undisturbed, layered sand was stable, so that there was no danger of the trench banks collapsing. There was only one case of a slide and it buried only the lower parts of the legs of one of his workmen.

One day Mr. R.R. Ross, who had designed the system, came to see how the job was progressing. However, at the time I did not realize his responsibility. He was always busy talking to the crewmen and unfortunately I did not interrupt. The following spring when I was working for the Public Health Engineering Division, Department of National Health and Welfare (DNH&W), he came to our office to visit Stan Copp, my supervisor. He told us that he was on his way to Fort Smith

Figure 185. *Water hydrant servicing people in homes that do not have indoor plumbing. Photo by Jack Grainge.*

to repair the water mains. He said when they turned the pumps on, water did not even reach the first house on the line. Obviously the pipes had broken. I am sorry that when Mr. Ross was in Fort Smith, I had failed to warn him of the problem. The error hurt his reputation. I apologized to him. We became good friends. He lived in Ottawa and we visited each other in our respective homes.

By the end of the summer, Mr. Ross, Water System Superintendent Bert Edge, and a workman completed the repairs on the mains. Then everyone in the settlement received treated water. Water was piped to the people living in the upper part of the settlement. All the people living in cabins in the lower part of the settlement obtained water from an insulated water hydrant.

Edge was hard-working, clever and practical. He was not a big man, but he was tough as he had shown many years before when he was a trapper. One day when his dogs and sled were going diagonally down a shallow slope, he ran alongside his heavily loaded sled, trying to hold it from tipping over. It did overturn. His dogs ran home leaving him with a broken leg. He crawled thirty kilometers through deep snow to his home in Fort Smith. It was extremely cold that night, and he had to continue crawling to keep from freezing to death.

During the summers, the ground under the pipeline from the pumphouse up the river bank slumped continuously, causing the pipe to break. Edge dug up the pipeline and repaired it. When the pipeline broke a second time, ingenious Edge connected the two ends of the pipe together with a surface, insulated and boxed, heavy rubber hose. To an experienced plumber that might sound like a simple solution. However for a person who had previously never worked with plumbing, not even having lived in a house with plumbing, that was a clever solution.

* * * *

Returning to my work in 1950, I stayed in Fort Smith to assist John Piard in taking river-bottom soundings for dredging around the boat dock at Bell Rock. Piard was then taking river-bottom soundings for dredging a harbor of refuge west of Fort Resolution. During storms on the lake, vessels could harbor there. Most often captains faced the storms in the lake with their barges in tow. The refuge was not maintained and in time silted up.

For much of his career, Piard was involved in taking soundings for the construction of wharves and dredging. The winding Slave and Mackenzie Rivers were continually gouging soil from their banks and depositing it in harbors and in sea lanes. Piard was a great companion, and after retirement we continue to meet to talk about old times.

While awaiting Piard, I walked about visiting local people. The weather was getting cold so I bought a beautiful, caribou parka for twenty-five dollars. I wondered why I should get such a bargain, until someone explained that it was made from winter pelts. The long hairs would soon begin to shed. Later in the winter this proved to be the case. In Edmonton, I used to obey the streetcar conductor and squeeze through the crowd to the back. I left a trail of caribou hairs on people's wool coats. In those days everybody wore a wool coat. In the office I had to hang my parka separately from other employees' coats.

In later years Keith Spencer told me about the waist-high, concrete monument at the Alberta-NWT border. On June 13, 1928, Edward Martin died. He had been a quiet, lone woodcutter supplying firewood for paddle wheel vessels plying the Slave River. S.H. Milton Martin, the local administrator, handled his $1500 estate. He placed advertisements in several Alberta newspapers, but received no answers. He therefore decided to use the money to erect this prominent monument in Martin's memory. Later a woman arrived in Fort Smith claiming to be Martin's widow. She demanded full payment of the estate, which Milton personally paid.

I enjoyed my time in Fort Smith, becoming acquainted with various people. One evening the people organized a community party. Everyone came, the civil servants, workmen, business men, wives and children, all visiting one another. Children ran about. There was no discrimination of any kind. Many people performed on stage, singing, telling jokes and performing skits. The women brought tasty food. What a wonderful evening!

* * * *

In about 1957, a company built a large hotel with a seven-meter deep septic system. My seepage tests showed that the sand would allow substantial seepage flow. However the seepage proved to be inadequate, so that the sewage had to be pumped out and hauled away.

In 1957, the Denny Logging Co. obtained a government loan to construct a plywood plant at Fort Fitzgerald. They built dozens of single-family houses in

Fort Smith for their future employees, but no manufacturing plant. The company went bankrupt. The houses were vacant for awhile but became occupied when the town's population increased.

The affable Paul Kaeser's retail business grew with the town. In 1950 he had a small store. He was a friend to everyone. Within a few years he built a larger store with a coffee counter. Not surprisingly that affable businessman became mayor and served many years in that capacity.

* * * *

Within seven years after the construction of the water plant, the population of Fort Smith grew from about 400 to 1000. It was the largest settlement in the NWT and was growing fast. At the time, no one thought that Yellowknife might supplant Fort Smith as the administrative center for the western NWT.

During spring floods, the water plant was not large enough to clarify the water for the growing town. Since a population explosion seemed to be in the offing, the district engineer recommended that the plant be either expanded or replaced.

DI&NA hired D.R. Stanley and Associates to design a sewerage system and a new, larger, water treatment plant. In 1958, Yukon Construction Co. of Edmonton built the systems. Bill Bales was the superintendent.

As a base for the water intake, Bales blasted a 1.7 meter hole in solid rock near the edge of the river. In this he set a poured-in-place caisson. He blasted a channel to deep water and laid a gravity water intake pipe in it.

He constructed a new, insulated pipeline up the high bank, supported above ground on timber blocks. Later the ground slumped, again at the midpoint and he inserted an additional pipe to relieve the tension.

A factory-made water treatment plant was shipped to Fort Smith in sections. It provided the same treatment as did the one replaced. The treated water flows into a reservoir below the floor of the plant. From here it is pumped to an insulated, steel, water tower.

In the meantime, the Government had supplied many residents with modern, wood-frame houses, with water reservoirs and sewage-holding tanks. The town then began trucked water delivery and sewage haulage systems.

* * * *

The sewers were 200 mm vitrified clay tile, leading to a sewage pump station at the east end of the Town. The minimum depth of soil cover on the sewers was 2.5 meters. Subsequently a sewer, constructed with only 1.8 meters of soil cover, froze at a manhole. I suggested that they try suspending an insulation platform immediately below the manhole cover.

A four-hectare sewage lagoon served the town. At that time that was standard for a town with a population of one thousand. The lagoon was located downstream of the town, seventy meters from the brow of the high river bank. The sand bottom was sprayed with a liquid that was supposed to prevent seepage.

Figure 186. *Sewage lagoon surrounded by sand. Although sand bottom was treated with a substance to block to interstices, the seepage prevented pollution of the receiving river for ten years until bottom growth blocked seepage and the lagoon overflowed. Photo by Jack Grainge.*

Luckily the sewage seeped away through the sand bottom, and thereby prevented pollution of the Dene fishing area immediately downstream. Seepage through sand removed all bacteria of human origin. For several years the sewage was nothing more than a puddle in the middle of the excavation. Eventually sediment plugged the interstices in the sand, and during the late 1960s the lagoon began to overflow to the river.

The DI&NA engineers added two quarter-hectare lagoons ahead of the main lagoon, which could be operated either in series or in parallel. The purpose of the small lagoons was to settle the solids from the incoming sewage. The solids could then be removed with a backhoe.

During the late 1950s, extensive research studies by the Sanitary Engineering Division, Alberta Department of Health and Welfare, showed that negligible treatment occurs in sewage lagoons during the winter. Therefore if people downstream might be affected by the effluent, lagoons should be large enough to retain the sewage for fourteen months.

The Town grew and the sewage flow increased. During the 1980s, the summer theoretical retention time had dropped to about three months. In winter when the ice was thick, it was only six weeks. People fishing downstream of the outlet complained about the discharge into the River.

The treatment did not conform with two sets of stream-pollution regulations, first, the Sewage Plumbing Regulations under the NWT Public Health Act, adopted during the 1960s, and second, the Pollution Control Regulations adopted by the NWT Water Board about 1979. Since the town population has now shrunk, the sewage flow has probably reduced.

Later, Louis Grimble became the Town's consulting engineer. He told me that the sewage should have flowed to a small lagoon west of the Town with the effluent discharging into the wetlands. However, when the lagoon was constructed, we engineers did not realize what excellent sewage treatment wetlands could provide.

Figure 187. *House built by Jacques Van Pelt has been designated a historic building; the window-encircled, circular crows-next is a good-fellowship meeting place. Photo by Jack Grainge.*

When Don Stanley was designing the system, I should have thought of suggesting that method of disposal. After all, the hospital and mission had been discharging their sewage into that swamp, with no adverse effect. There was an odor at the disposal point but it was not noticeable fifteen meters away.

During the mid 1960s, Bill Parker, Engineer in the Edmonton office of the DOT, constructed a sewage lagoon at the airport, six kilometers downstream of Fort Smith. It served the house for the airport maintenance workmen and the airport waiting room. The lagoon was small, approximately 11 m x 11 m, but it was large considering the small population it served. Parker provided a plastic lining that allowed the lagoon to fill. I did not see it, but Mike Pich, of our office, reported on it. Fortunately it did not overflow to the river during my career. Providing a liner was a mistake, but most likely there were leaks in the lining that allowed seepage into the sand bed. That leakage prevented pollution of the river.

*   *   *   *

During the 1960s, Al Smith, Engineer for NTCL, asked me to suggest a summer sewage disposal system to serve the NTCL office and workmen's quarters at Half Way. To avoid contamination of the river immediately upstream of the Fort Smith water intake, we decided there should be no overflow. We designed a system consisting of a septic tank discharging into an oversized lagoon with no overflow, depending entirely on evaporation and seepage into the almost water-tight, clay subsoil. The system worked well.

Eventually the NTCL moved that personnel camp to their loading dock at Bell Rock, 12 km downstream of the town. The kitchen and bunkhouses had running water hauled from hydrants in Fort Smith. However sewage disposal was primitive. During the mid 1960s, Monte Stout, who had replaced the late Al Smith, asked me to suggest the design of a summer sewage disposal system. I suggested a septic tank system with a shallow seepage field pipe. I did not see the system, but he told me that it operated well.

* * * *

In 1960 Dr. Falconer, Regional Director, Medical Service Directorate, DNH&W, asked me to accompany a doctor to Wood Buffalo National Park, southwest of Fort Smith. We were to make sure that the wildlife officers would not contract anthrax from the bison. Anthrax is fatal to all mammals, although carnivores, including man are somewhat resistant.

To go back in history, wood bison had been there for centuries. In 1902 Dr. Robert Bell, Dept. of Mines and Technical Surveys, hired Charles Camsell to survey the source of the salt used throughout the NWT. He and a helper, Duncan McKay, paddled up Salt River, 32 km downstream of Fort Smith, and found a salt spring. The salt-saturated water was evaporating, leaving salt on the ground.

Figure 188. *Bison in Wood Buffalo National Park. Photo by Robert J. Hudson.*

Their second chore was to make a survey of the wood bison that had survived in this area. At one time there had been thousands of bison extending as far west as Fort Nelson, B.C. and north to Great Slave Lake.

On horseback Camsell found only a small herd near Flat Grass Lake. Based on their inquiries among Dene, they concluded that there was one herd between Salt River and Peace Point, and another one near the watershed of the Nyarling River. In the whole area they thought there were about three hundred bison.

During the years 1925 to 1928, the federal government moved 6673 plains bison, by rail and barge from Bison Recovery National Park at Wainwright, Alberta to Wood Buffalo National Park where they bred with some of the wood bison. At present there are about 2100 plains bison in the park, with another 600 in the Slave River lowlands.

In 1951, Dr. B.I. Love, Superintendent of Elk Island National Park, told me that the plains buffalo shipped there had brucellosis and other diseases. Therefore he had recommended that all the animals in the affected part of the herd be destroyed. He said he could replace them with disease-free buffalo from Elk Island National Park. He thought that they would multiply satisfactorily. His recommendation was not accepted.

In early 1960, Dr. Lawrence Willoughby, the local federal medical doctor, diagnosed anthrax in a wildlife officer. A few days previously, the officer had butchered a sick buffalo. From then on the wildlife officers have been trying to prevent the disease from spreading among the herd. They buried dead buffalo together with liberal coverings of lime. Unfortunately, many of the bodies were not found. Furthermore, spore forms of the bacteria, dripped from the mouths of sick animals before they die, remain in the soil for many years and can be ingested or inhaled by other bison.

During the 1960s, park officials established a herd of wood buffalo from the northern part of Wood Buffalo National Park in the southern part of Elk Island National Park. They are completely isolated from the plains buffalo in the northern part of that park. Dr. Claude Rouget found that some of them were infected with tuberculosis. Therefore they destroyed all the mature animals and started a new herd from calves that were free of the disease.

Dr. Rouget vaccinated for anthrax two thousand of the plains bison in Wood Buffalo National Park. He told me they are too wild to to be properly herded into corrals for vaccination. The wildlife officers chase the bison by helicopter into corrals, but many of them become trampled and killed in the herding process.

* * * *

During 1967, Canada's hundredth year, Yellowknife became the capital of the NWT. The ever-optimistic Kaeser hurried to Ottawa to save his town. He talked the powers into making Fort Smith the administrative center for a huge region containing half the population of the Mackenzie Region, including Yellowknife and all the towns north to Wrigley. However, years later this regional office was transferred to Yellowknife.

References:

Grade 6 class. 1979. *On the Banks of the Slave.* Yellowknife: GNWT Department of Education, Teachers and students at the local elementary school, under the direction of Joseph Burr Tyrrel and Dennis G. Siemens.

Camsell, C. 1954. *Son of the North.* New York, N.Y.: D. McKay Co.

Dusel, F. 1995. Personal communication.

Edwards, R. 1995. Personal communication.

Leising, Fr. W.A., O.M.I. 1995. *Arctic Wings.* Garden City, N.Y.: Doubleday.

Piard, J. 1995. Personal communication.

Rouget, C. 1995. Personal communication.

Spencer, K. 1995. Personal communication.

Stout, M. 1995. Personal communication.

DISTANT EARLY WARNING LINE

During 1956 and 1957, I witnessed a part of a mind boggling drama in the Arctic. In the name of military emergency, ordinary men, facing almost impossible Arctic weather, accomplished one of the greatest of construction feats in the Far North.

By 1953, Russian low-flying, long-range bombers could cross the roof of the world and drop atomic bombs on American and Canadian cities. The radar stations of the Pine Tree and Mid-Canada Lines were unable to detect low-flying planes soon enough to protect these cities. We desperately needed a far-north, radar warning system. Quick!

Figure 190. *Radar aerial dome at Tuktoyaktuk. Vapor trails of fighter planes that leave a station in the USA in the morning and return that evening. They were refueled in the air as they passed over Edmonton. A cousin of mine was one of the American crew who were stationed in Edmonton. Photo by Jack Grainge.*

The Joint Defense Board of Canada and United States decided that we needed a chain of radar stations along the north coast of Alaska and mainland Canada and the south coasts of some of the more southerly Canadian islands. Subsequently the Distant Early Warning Line, the DEW Line, as it became popularly known, was extended westward along the Aleutian Islands and eastward across Greenland and Iceland.

Engineers and architects of Western Electric Co. designed and supervised the construction of the DEW Line. After construction, the Federal Electric Co., a

Figure 191. *Norm Lawrence between Canadian and US flags flying at Cambridge Bay station at high sun January 12. Photo by Jack Grainge.*

subsidiary of Western Electric Co. took over the operation of the Line. The RCAF assumed supervision of the Canadian section.

The American army built the Alaskan section. They supplied these stations by ships from the USA and in winter by Cat trains from Fairbanks. Foundation Engineering Co. of Canada constructed the eastern Canadian section. Ships from Montreal and the USA and planes from Iqaluit, supplied this section.

Northern Construction Co., a Canadian subsidiary of Morrison Knutson Inc. and James W. Stewart, built the western Canadian section. This section extended from a few kilometers east of the Alaska-Yukon Territory border to Shepherd's Bay, on the west coast of Boothia Peninsula. There were twenty-four stations in this section. There would have been twenty-five, but the planners decided that the westernmost site in Canada should be constructed by the American army. It was less than ten kilometers east of the border.

* * * *

During September 1956, I examined the water supply and sewage systems of the stations. I arrived at one of the stations a year and a half after the start of construction. I expected to find the work only half done and to hear that delays were due to bad weather and slow delivery of equipment. Such explanations for jobs unfinished on time were common in the North. However the construction was largely completed. Also key workmen on practically finished stations were transferring to another huge construction job.

Fortunately for me, at one of the stations I met an inspection engineer representing Western Electric Co. He described the design and construction of the DEW Line. In the planning stage, the U.S. Army first built prototype stations, one at the design headquarters in Paramiss, New Jersey, and another on Barter

Island near the north coast of Alaska. The planning engineers and support personnel worked and lived in these stations, making certain that all equipment would operate satisfactorily, and that the operators and workmen would enjoy comfortable living conditions.

All stations consisted of 8.5 m x 4.9 m independent units, called modules. The modules strung end to end formed trains. They were spaced 15 cm apart and connected by enclosed doorways. They were heated with hot air from the diesel-electric engines, and in emergencies by electric heaters.

Figure 192. *Uninsulated aluminum pipeline carrying warm sewage from Cambridge Bay Station to a small lagoon without freezing in winter. Photo by Jack Grainge.*

If a fire occurred in one unit, it could be bulldozed sideways away from the train. Also the builders used the most up-to-date of fire resistant materials and paint, and mounted a fire extinguisher in each unit.

Six of the radar stations were main stations, well spaced from Alaska to Baffin Island. They radioed messages from the sites to NORAD. Each main station consisted of two, 25-unit parallel trains spaced about 25 meters apart. An enclosed overpass, high enough to clear maintenance trucks, connected them. If one train burned down, the other train, a backup, would still be operational.

Three or four auxiliary radar stations, consisting of eight-module trains, were located approximately 160 km apart, between the main stations. They were to gather information about incoming enemy flights and report it to the main stations.

Unmanned intermediate stations were located approximately midway between the auxiliary radar stations. They relayed radio transmissions between the radar stations. These five-module trains contained living quarters for visiting repair crews.

Each auxiliary station contained a diesel-electric power plant, operations equipment and office in two modules at one end of the train. The next modules contained a kitchen, dining room, living room, washroom and individual bedrooms. The end module of the train contained water treatment equipment, and water and sewage storage tanks and pumps.

The building foundations, roads and airstrips consisted of gravel pads, one to one and a half meters thick. The trains were aligned in the direction of the prevailing winds and constructed on one-meter-high, open frameworks. The winds blew mainly around, and to a small extent under, the trains. The wind underneath may have reduced snow drifting but huge drifts formed around the porches.

Figure 193. *Water tank truck being filled, Gladman Point. Photo by Jack Grainge.*

* * * *

I was surprised to find that fresh fruit, vegetables, meat, eggs, and milk arrived daily by airplane. The milk was pasteurized at a plant in California at a much higher than usual pasteurizing temperature. At that time high-temperature pasteurization of milk was a new process, not then used by Canadian dairies. At room temperature, the milk retained its freshness for a month. However during the first few months of the construction, the men had used powdered milk.

The cooks on the Line were master chefs. At almost every station the workmen boasted that they had "the best cook on the Line." For example one evening, a few workmen and I were enjoying a recent movie. While the projector operator was changing a film, the cook came in serving some newly baked cookies and cakes.

The recreation unit contained a bookshelf with books, cards and board games. It was enough recreation for construction men who worked ten-hour days, seven days a week.

The water treatment systems provided coagulation and turbidity removal plus chlorination. In all cases that I remember, the water was clear so that only filtration and chlorination was required. A storage tank at each site held sewage at room temperature to be hauled away as required. In all cases the equipment was operated well.

I made a suggestion that probably saved millions of dollars. They had planned to truck the sewage away from the stations. I pointed out that they could pipe the sewage to nearby low areas. Sewage lagoons would develop in these places and provide adequate treatment. Even if the sewage entered the sea directly, no problem would result. In fact the western Canadian arctic seas are nutrient-poor, and the introduction of sewage would slightly benefit the ecosystems.

The systems operated well. In winter, the room-temperature sewage flowing from the large storage tank at each station warmed the discharge pipe. Providing the pipe had no dips, it drained before becoming cold.

In the Western and Central Canadian Arctic, the sites were located on frozen flatland. In all cases garbage was heaped on the ground surface a distance away from the airstrip and buildings. Construction wastes along with unneeded oil barrels were piled separately, convenient for the resourceful Inuit who built one- or two-family villages nearby.

The site at Tuk was the only one in the western section which was near an existing settlement. Fraternizing by workmen with local people was not allowed. Because of the swampland around the tiny community of Tuk, the garbage could not be transported a long distance away.

Construction of the DEW Line took place more than a decade before PCBs were found to be dangerous to the environment. I am confident that scientists will discover a simple, organic, method of reducing PCBs to harmless components.

* * * *

Before the DEW Line construction, the only passenger service to the high arctic settlements was by way of small planes fitted with skis in winter and pontoons in summer. They landed on either the sea or nearby lakes.

During the summer of 1954, survey crews in small, single-engine, pontoon-equipped planes, examined and reported on the possible sites for the stations.

Northern Construction Co. started with the building of an airport and an auxiliary radar station at Tuk. To start supplying other stations they hired all available bush pilots together with their small planes. They saved time by landing men and equipment on nearby lakes, which froze earlier than the sea. When ice had formed on the sea they flew in garden-size bulldozers, sleds and equipment on Fairchild, C119 planes, known at the time as flying boxcars. These planes skimmed along low over the ice, dropping the loads, a piece at a time out of their open rear doors. To slow the landings of the freight the pieces had parachutes on their tail ends.

Using these small bulldozers, they cleared snow off large sheets of ice to allow faster freezing. Eventually they had runways suitable for the landing of Fairchild C119 planes. With these, they brought in D8 bulldozers, huge trucks, diesel-electric generators and finished wall, floor and roof sections of the modules.

Freighting to the Western Canadian Section required careful planning. Shipping was by both air and water and by combinations of the two. NTCL hauled freight by truck to Hay River, by barge across Great Slave Lake and down the Mackenzie River to Tuk and by ship on the Arctic seas. NTCL hired frogmen to check the depths of the water near the unloading points. The shipping season on the Mackenzie River and the seas in the west is four months long. In the central Arctic, the seas are open for two months. The shipping season to Yellowknife is five months long, so some freight was barged there from Hay River and then flown to the sites. Much freight was flown all the way to the sites from Edmonton. It must have taken a team of whiz dispatchers to fit it all together.

Figure 194. *Pressure ridge in sea near Cambridge Bay. Meltwater from the ice near the top of the ridge is salt-free. Photo by Jack Grainge.*

Ships of the American Navy brought freight to coastal points by way of Bering Strait. They also carried gasoline and diesel oil in their ballast tanks.

The western Canadian section was operational within two years after the start of construction. Construction of the eastern section, with some sections built in difficult mountainous locations, took a year longer.

* * * *

Herb Donaldson was the project manager of the western Canadian section of the Line. He was a key organizer and planner of the work at all twenty-four stations. He answered all my questions, seemingly aware of everything down to the smallest details. He was a congenial, mild-mannered gentleman, highly respected by his assistant, Percy Simpson, and all the station superintendents. I met him several times as he traveled from site to site, checking with the station superintendents.

Figure 195. *Young Point, on mainland coast north of Coppermine. Huts made from packing cases are better than snow houses; in midwinter they are covered with snow. Parents like to see the children enjoying themselves. Photo by Jack Grainge.*

The construction required many unusual and interesting methods. The camps were dry, that is, no liquor was allowed. Many alcoholics joined the labor force to dry out and perhaps cure themselves. At the end of six-month terms of service, all workmen received leave with paid airfare to Edmonton. Many men went no farther than Edmonton. They had worked ten hours per day and seven days per week at the highest rates of pay for tradesmen, with no place to spend their money. Some could not resist spending their seemingly inexhaustible wealth on parties in their hotel rooms. Within weeks some of them were broke and therefore signed up for another term on the DEW Line.

Northern Construction Co. kept savings accounts for all workmen. Most workers saved their money and after serving successive terms, many were able to set up their own construction, sales or service companies. Marty Larson, an accountant, left a bank to earn $500 per month plus 8% holiday pay. He soon saved $5000, a small fortune for a young fellow in 1956-57. At a main station, Pin Main, on Cape Parry, on a monthly salary he worked some days as long as eighteen hours. He earned every penny he made. Station foremen were paid $2.95 per hour and laborers $1.20 per hour.

Larson arrived early in 1956. He barely escaped freezing night after night in a four man, insulated tent, with a plywood floor, and barely warmed by an oil heater. He recalls many days when the temperature hovered below -40° C with the wind 120 kilometers per hour.

Larsen learned the merit of the no-liquor rule. The American personnel were able to bring overproof rum on the sites. Some of the workmen and single Inuit joined an American in celebrating Christmas and became depressed. Even without the liquor, a few men became bushed.

Donaldson hired as many Dene and Inuit employees as he could place. These men became truck drivers, bulldozer operators, handymen and laborers. Some Inuit workmen brought their families to the sites. Married Inuit lived a kilometer or so distant from the stations. At the wastes sites they salvaged wood packing cases, hauling them to their camps by dog team. With wood sections and pieces of glass they built huts. These huts, although tiny, were warmer and more spacious than tents and snow houses.

In early winter, after snow drifts had accumulated, they piled snow blocks around the huts to make them both warmer and draft-free. Some of them constructed insulated double roofs. Using snow blocks, a few built tunnel entrances to the doorways.

Figure 196. *Beautiful mother and child at DEW Line station, Gladman Point, King William Island. Photo by Jack Grainge.*

Most of the Dene workmen could speak English, but in the western Canadian section of the DEW Line, the Inuit spoke three different dialects of Inuktitut and few could speak English. The Company hired three experienced northerners to advise the foremen on dealing with the Inuit and Dene. Bert Boxer, a former trapper and workman at Aklavik had worked many years with Inuit. Father Meteyer, was a young Catholic priest from France who had served at Inuit settlements. He was both an interpreter and a chaplain. Subsequently he wrote an interesting book,*I, Nuligak,* an interpretation of an Inuk's life story, a translation of the life story

Figure 197. *Family with sled made from discarded packing cases, Lady Franklin Point. Photo by Jack Grainge.*

of Nuligak, a smart, tough Inuk. Jamie Bond was a university graduate. He had done his best to learn one dialect of the Inuit. However it seemed to me that the superintendents could make their wishes known to Inuit without the help of an interpreter.

* * * *

Besides keeping a watchful eye out for intruding, foreign planes, the DEW Line radar operators provided surveillance of all domestic flights. The staff of the stations allowed private airlines to use the runways and provided needed assistance to travelers, but not accommodation. During 1956, a boy was sick at Taloyoak. Herb Donaldson arranged transportation for Nurse Kay Dier on a regular flight to the nearest station to Taloyoak, and then by special plane to Taloyoak to pick up a sick Inuit boy and brought them to Tuk, for transportation to Edmonton.

Figure 198. *Children on sledge at Radar Site, Young Point. Photo by Jack Grainge.*

During construction, generally women were barred from the DEW Line stations. In March 1956, Dr. Bill Davies and Nurse

Winnie Rosco, on a private plane flying from Tuk to Coppermine, were forced by weather to land at one of the stations. Winnie does not know which one. The station superintendent invited them to dinner. A waiter escorted her to the washroom door and ensured privacy while she was there.

Figure 199. *George Porter, HBC Manager and old man who danced at a reception for Governor General Vincent Massey at Gladman Point, when he visited the North. Photo by Jack Grainge.*

Winnie heard an announcement on the P.A. system warning workmen not to use the outdoor facilities while the visiting party was there. The outdoor facilities were urinals, consisting of empty, oil barrels with part cut down to a convenient height. In winter they used a three-sided, waist-high, snow-block enclosure.

Until the washrooms in the modules were finished, temporary privies consisted of a row of six holes in an elevated seat inside a module. Empty oil barrels under the seats could be removed and replaced by way of a temporary outside access door.

* * * *

In the late 1960s, due to improvements in communications systems, the Federal Electric Co. abandoned the Intermediate stations. In 1993 they also abandoned some of the auxiliary stations and converted the rest to unmanned stations. They added others so that there were thirty-six all together. At the end of the cold war, all of these small stations were abandoned, leaving only the six main stations. Since the abandoned auxiliary stations have excellent living quarters, I hope that tourist entrepreneurs will take over some of them. They are monuments to an amazing arctic construction feat. Some of them should be preserved.

Indirectly, the cold war with Russia provided another impetus for change in the North.

Figure 200. *Residences for families of two permanent Inuit employees. Photo by Jack Grainge.*

During the two years 1955-57, the American Army changed the face of transportation for communities in arctic Canada. Until then people depended on small planes equipped with skis in winter and pontoons in summer. During in-between seasons when rivers, lakes and seas were freezing or the ice was breaking up, access to the communities had been impossible.

Soon after the beginning of the construction of the DEW Line, Canada had a chain of well-constructed, all-weather, gravel airstrips strung approximately eighty kilometers apart, all along its north coast. The DEW Line personnel provided navigational and emergency assistance to travelers. These airports have assisted the mining, oil and tourist industries, as well as the development of the arctic communities.

I will never forget the two, two-week visits I spent along the DEW Line. The careful overall and detailed planning, the efficient organization of the construction, the close cooperation between management and workers, and the delicious meals and excellent entertainment facilities and accommodations are highlights in my memory.

References:

Berube, R.A. 1968. Report of Avcon Consultants Ltd.

Larsen, M. 1998. Personal communication.

McNicol, D. (Major General Don). 1998. Personal communication, Retired, RCAF.

Spraggs, J. 1999. Personal communication.

SWIMMING POOLS

For centuries the Inuit and NWT Dene spent most of their working time on the rivers, lakes and seas. Every year a score or more of them drowned because they could not swim. They sank even when within one stroke of safety–someone's hand, a boat or the edge of ice. Because the lakes, rivers and seas were cold, they never learned to swim.

While everyone in the North recognized the problem, Father Max Ruyant, in charge of the Catholic school dormitory in Inuvik, was the first to come up with a solution. In the late 1950s he built a heated-water, polyethylene-lined, plywood pool in a shed. With the local residents he formed a transport company to bring in food supplies. The profits funded the construction of an ice arena and the indoor swimming pool.

In the 1960s a swimming program began at Pine Lake, ninety kilometers south of Fort Smith. However the bus rides were too long, and the time that the pupils could withstand the 13°C water was too short.

In 1963 ball-of-fire Gil Gilmet, Director of Alberta-NWT Red Cross Water Safety Division, set his sights on a territory-wide program. He and Jacques Van Pelt, Director of Recreation Division, NWT, organized a course for instructors of swimming and water safety. They realized too that swimming would be an activity to reduce youth delinquency.

Gilmet recommended the enthusiastic Walter Scott to Mark De Weerdt, President of the Yellowknife Red Cross, to conduct a swimming instruction program. In 1964 Scott started a program at Frame Lake, a small lake that becomes warm on sunny, summer days. Scott had minimal funds for living. He managed in a shack using a kerosene stove and a sleeping bag. His enthusiasm was so contagious that soon he became overwhelmed with 275 registered students. He quit. When Van Pelt provided funds to hire assistant instructors, Scott began again. He continued with the program through the summers of 1965 and 1966.

In 1966 Gilmet, Van Pelt, Dr. Murray Smith, Professor of Physical Education, U of A, and Scott organized a meeting of potential leaders to plan programs to serve several communities. At the meeting they discussed a program then being tried in North Carolina. Polyethylene-lined, portable pools were hauled on flat bed trailers to districts populated mainly by low-income families.

From necessity economy prevailed. A typical scenario would include a heated, polyethylene-lined, plywood pool in a shelter of some kind—an existing, unused building, or an inflatable, polyethylene bubble. Water treatment equipment would consist of surface skimmers, a recirculation pump, a water heater, a filter and a chlorinator. There would be a wooden walk from the pool to two dressing rooms, a shower and soap. The instructor needed to be knowledgeable, enthusiastic and dedicated. Usually the right person was an underpaid university student seeking adventure in the North.

In letters to community councils, Gilmet explained the organization and the program. The communities must provide accommodation for the instructor, a building to house the pool, change rooms, the lumber and polyethylene lining, the equipment and the labor to build the pool. Van Pelt's Recreation Division matched funds that the communities raised. The communities held dances, bingos, raffles, and solicited donations—anything to get the much-needed swimming courses.

During the summer of 1966, Van Pelt and Gilmet experimented with an 8.2 meter diameter, prefabricated, polyethylene-lined pool in an arena at Fort Simpson.

In May 1967 Scott supervised students at Grandin College, Fort Smith, in building a rectangular, polyethylene-lined, pool in a Quonset hut. Then he returned to continue his program at Yellowknife. In 1968 Scott moved the pool, equipment and program from Grandin College to Pine Point.

Each spring Gilmet organized meetings of everyone involved, including the university students who would be both the swimming instructors and foremen for the construction of the pool, housing and the plumbing equipment. Scott, the field supervisor, would visit the sites to help the instructors and ensure that the organizations were running smoothly. If anything was lacking, the instructors were to phone Gilmet, who would ship the item on the next plane north.

Gilmet and Professor Don Smith, Chairman of Red Cross Water Safety, Alberta-NWT, answered instructors questions by phone or letter. Also they conducted clinics at swimming pools to which they could drive. All of Smith's eight children became outstanding competitive swimmers. At the 1978 Commonwealth Games, Graham Smith won eight medals.

The pools did not meet minimum sanitation standards. Teaching people to swim and save lives overrode those shortcomings. Well-chlorinated water made up for the lack of good showers. Well-dried feet and foot powder between toes helped in preventing athletes' foot. If foot powder is not available, they could use flour.

Our office supplied test kits to each instructor for determining the concentration of dissolved chlorine and the pH (acid-base level) of the water. Frank Gillis, Environmental Health Officer, DNH&W, Yellowknife and his successor, George Woodget, checked the testing by the operators. Later John Shaw, Engineer, of our office also helped.

We discovered a surprising side effect. Swimming in chlorinated water cured skin diseases that previously had plagued many of the children.

Instructors constructed the pools to fit the buildings in which they were housed. In 1969 Tony Bouwmeester, the instructor at Inuvik, constructed a pool in a curling rink. He set the pool within two lanes, approximately 18 m x 8.5 m. Later that summer he went to Norman Wells to set up another pool. It was a manufactured pool, size 12.2 m x 6.1 m x 1 m depth of water, with steel sides, a fiberboard bottom and polyethylene liner.

In 1968 NTCL offered Gilmet a covered, thirty-seven meter barge with a twenty-five meter enclosed area. It could serve communities that could not afford a pool. Gilmet and I met at a firemen's convention in Hay River where I was guest speaker. The next day we waded through hip-deep snow drifts to peek through the windows of the barge. We agreed the roof could be raised for a polyethylene-lined pool. A $5000 donation by the Eric Harvey Foundation, Calgary, paid the bill. The pool was 11 m x 5.5 m x 1.1 m deep.

Gilmet organized pool construction and swimming programs at Yellowknife, Hay River, Inuvik, Fort Smith, Pine Point, Talita, Fort McPherson, Fort Simpson, Tuktoyaktuk, Fort Churchill, Baker Lake and Rankin Inlet. The pool on the barge served Fort Providence, Fort Good Hope, Fort McPherson, Aklavik and Tuktoyaktuk.

In approximately 1968, an enterprising company built a multi-storey apartment building in Yellowknife. They constructed a substantial swimming pool on the main floor. Yellowknife's Red Cross program was transferred to this pool.

Accidents occurred at Fort Smith and Inuvik when the bracing at the lower parts of the pools gave way. The outward hydraulic pressure at the base of a pool is substantial.

In 1970 I met Roland Gosselin at the organizational meeting in Yellowknife. At the time he was the physical training instructor at Inuvik. He worked with the Inuvik swimming program. In 1974 he became the NWT physical education consultant based in Yellowknife. He worked extensively with the Red Cross program. Also the Canadian Coast Guard began providing useful assistance.

I realize now that I witnessed the evolution of a much needed swimming, water-safety and lifesaving program. The far-sighted Gil Gilmet, the rightest of men in the right place at the right time, provided the leadership. Walter Scott was reliable in the development, and Jacques Van Pelt came through with funds and guidance. Gillis, Woodget, Shaw and I played minor roles. We praised the organizers for realizing that the objective was too important to be thwarted by too-carefully following difficult sanitation regulations. We suggested simple, alternative practical ways to prevent the spread of diseases and infections.

Following the 1970 workshop in Yellowknife, I was pleasantly surprised when Brian Bailey, representing the City of Yellowknife, presented me along with Gilmet and three others, Arctic Adventurers' Certificates.

After the workshop John Shaw and I wrote a twenty-two page manual containing advice regarding the construction and operation of the pools. Gilmet told me that he gave copies to all of the operators and helpers. It became everybody's bible.

Since my retirement, the imaginative Roland Gosselin expanded the program to meet the needs of the growing towns. A self-taught handyman in several trades and professions, he planned fourteen-meter by seven-meter permanent pools in insulated and heated buildings. By so doing he has doubled both the hours per day and the length of the swimming season. He initiated watersafety campaigns which prompted the Canadian Coast Guard Service personnel to organize courses in water boat safety.

References:

Bouwmeester, T. 1997. Personal communication.

Public Health Engineers Division. 1970. *Manual on Non-permanent Swimming Pools in the Northwest Territories.* GNWT: Department of National Health and Welfare.

Gilmet, G. 1997. Personal communication.

Gosselin, R. 1997. Personal communication.

Scott, W. 1970. *Report on NWT Portable Pool Workshop, Yellowknife.* M.A. Thesis: Faculty of Education and Recreation. University of Alberta.

COMMUNITY ENVIRONMENTAL ENGINEERING IN SCANDINAVIAN COUNTRIES

The capitals of Norway, Sweden and Finland are near sixty degrees north, and most of the land of those countries is farther north. Greenland is entirely north of the sixtieth and Iceland is just south of the Arctic Circle. Thus for centuries Scandinavians have learned how to plan communities to cope with long winters, with short days and a low sun.

I needed to see examples of how planners and engineers in other northern countries solved community environment engineering problems. I applied to the World Health Organization for, and was granted, a travel fellowship for study in the Scandinavian countries. In September 1968 I traveled in northern Norway, Sweden and Finland, and in June 1969 in West Greenland and Iceland.

I prepared reports after each of the two trips. I forwarded these to various government and consultants' offices, hoping that the planners would get ideas from them. We in our office used the ideas in making our reports on communities.

A residential subdivision, then being planned for Reykjavik, Iceland, exemplifies some of these principals. It occupied a south-facing slope of a hill, and light industries occupy the north-facing slope. In the residential subdivision, high apartment buildings ranged along the top ridge of the slope, and successively lower houses along east-west streets down the slope. The front rooms of as many houses as possible would have view windows facing south.

Figure 201. *Plan of community to be built on south-facing slope in Reykjavik, Iceland. Photo by Jack Grainge.*

The hill, surmounted by high buildings protect the residential district from cold, north winds. The south-facing slope results in the sun's rays striking the ground more directly and therefore being warmer.

Kautokeino, a Lapp town in northern Norway, was being moved from its original north-facing site on the south side of a valley to a south-facing site on the opposite side.

A town center was being planned for Alta, a town of six thousand people, in northern Norway. The planner took steps to ensure that the town center would be busy during both days and evenings. A pedestrian street runs between the fronts of sixty-six buildings arranged in two rows. A road runs around behind them. The buildings are four, six and eight stories, with stores and warehouses on the main floor, offices on the second floor and suites above. A junior school and a gymnasium for the whole town are nearby.

Figure 202. *Graveyard located on unserviceable lowland, considered so because a sewage pump station would be required (Succurtoppen, Greenland). Photo by Jack Grainge.*

Ralph Erskine planned many interesting buildings to keep northern towns compact. Two apartment buildings he designed for downtown Kiruna, Sweden, have stores on the street level, offices on the second level and employees of the stores and offices occupying some of the suites above. In Sweden restaurants occupy the top floors of many high buildings.

The planner who designed Tapiola, twenty-five kilometers from Helsinki, incorporated many of these principles. He transformed land, thought to be a worthless rock hill, into what they call a garden city. He located high apartment buildings on the hill, and a restaurant is on the top floor of a high public office building. Sports fields, that do not require sewers, occupy the nearby low land. He planned the community carefully so that no sewage pumping station was required.

Scandinavians are sportsmen. Almost half the families in Finland have boats. Most towns had indoor sports centers and outdoor soccer fields. There are many indoor swimming pools and Finnish saunas near the washrooms.

* * * *

In many cities, the householder owns only the land occupied by his buildings, thus engineers can locate water mains and sewers within a meter or two of buildings. This reduces freezing of service pipes, the part of the systems most vulnerable to freezing.

In towns in Scandinavia where water mains and sewers are vulnerable to freezing, they are buried in the same trench. The water mains are separated about twenty centimeters horizontally from the sewer mains, and their undersides are twenty centimeters higher than the tops of the sewers. In Sweden a thirty centimeter wide, colored plastic ribbon is laid in the ground twenty centimeters above the highest pipe. This warns a workman who is digging that he is near the pipe.

Figure 203. *Sod house in Upernavik, substantial, but I saw only two–this one and another in Eggedesminde. Photo by Jack Grainge.*

Planners for northern Scandinavia and Greenland make towns compact so that, regardless of freezing weather, piped water and sewer systems are feasible. In northern Canada, spread-out communities require water and sewage haulage tanks. Such systems result in costs of up to forty cents for a toilet flush. Also water tanks are never clean, and sewage tanks in heated crawl spaces below buildings commonly overflow. Sewage tanks buried outside often freeze.

At the time of my visit, the toilets used throughout Scandinavia used much less water than those used in the United States and Canada. The toilet is flushed by a pull on a rod rising up through the cover of the flush tank. Pressing the rod down before the flush tank has emptied can reduce a flush further. Recently I installed a toilet in my house which uses less than a quarter as much water as a standard flush toilet.

*   *   *   *

In June 1969 I visited the large communities in Greenland from Godthaab, now called Nuuk, the administrative center, lat. 64° N to Upernavik, lat. 73° N. Upon my recommendation, in June 1977, I, along with several other Canadian engineers, visited Greenland towns from Sonderstromfjord to Jacobshavn.

Figure 204. *Greenlander in her Sunday Best. Photo by Jack Grainge.*

Before World War II, the Danes tried to protect the Greenlanders (Inuit) from the influences of modern civilization. They allowed a few missionaries, medical personnel, explorers and fur traders into the settlements, but tried to prevent other contacts. During the war the American army occupied strategic places, and hired local people. These people liked the new lifestyles, and in post-war years demanded modernization. In response the Danes began building houses and schools in the larger settlements. In 1950 the King and Queen visited some of the settlements and publicly deplored the primitive living conditions.

In response the Danes began developing towns in Greenland that they expected to become economically viable. They decided that (1) all new housing should have piped water-and-sewer systems. (2) People moving to these developing towns should live in apartment buildings. (3) The first two years in school should be taught in Greenlandic. (4) In the elementary grades Greenlanders should attend schools in both Greenland and Denmark. (5) Greenlanders should have vocational training.

Greenlanders trained as social workers played an important role in helping their people in adjusting from a subsistence life of hunting, trapping and fishing to a wage economy in industrial communities. The high level of literacy among Greenlanders, almost 100%, and the high level of training of tradesmen are commendable.

Although racial relations were often not harmonious, integration of Greenlanders and Danes is encouraged by not allowing residential ghettos of either group to develop. Apartments are separated by concrete walls and floors that are fire-resistant and sound-attenuating. As well members of both races play sports together.

I thought that interracial problems experienced in Greenland were the same as those we were facing in northern Canada. However, they handled them better.

Ships from Denmark can reach all communities in Greenland. Also the climate is warmer than that of arctic Canada, with all Greenland towns to be developed being in the

Figure 205. *Apartment houses set so that sewers drain from one to another (Jacobshavn, Greenland). Photo by Jack Grainge.*

Figure 206. *The sewer line from the apartments then drains to the ocean. Photo by Jack Grainge.*

subarctic region. In other respects the physical problems in Greenland are more difficult than those in Canada.

To reduce heat loss, water mains and sewers are laid as close to buildings as physically possible. Diesel-electric generators provide the power. In Godthab waste heat from the generators is used to heat pipelines and apartments. Plans were being made to do so in other communities.

All communities are situated where gravity sewers without pump stations can serve them. Excluding Sondrestromfjord, which consists of a hotel for transients and an American military base, there are only two sewage pump stations in Greenland. One serves Sukkertoppen, which is on an island, and the other Julianehaab, an old town. Storm water drains by systems of paved open channels. The Greenlanders have one important advantage. Their heavy rainfall and snowfall result in much of the winter accumulation of pollution being washed into and down the road gutters. In northern Canada precipitation is low, so that winter's wastes remain in the settlements. Thus, safe disposal is more difficult.

Since 1970, international conferences for people involved in technical work in north polar regions have been taking place. From these, the best systems should evolve.

References

Grainge, J. 1968. *Environmental Engineering in Northern Scandinavia.* Edmonton: Public Health Engineering Division, Department of National Health and Welfare.

Grainge, J. 1969. *Environmental Engineering in Greenland and Iceland.* Edmonton: Public Health Engineering Division, Department of National Health and Welfare.

EPILOGUE

In the early 1950s when I first went north, most Inuit and Dene were living in groups of a few families. Most of them followed their traditional life styles. Children's education involved learning to hunt, fish, trap, cook, sew, build primitive but ingenious homes and travel long distances by dog team. Many children attended residential schools in various communities.

At the end of my career, twenty-seven years later, the population of the Northwest Territories had tripled. The people were living largely in communities, all of which were accessible by telephone, radio, TV and scheduled airline flights. There were roads to many of them and a railroad to Hay River and Pine Point. Since my retirement, the branch of the railway to Pine Point has been abandoned.

With ever increasing advancements in planning, manufactured products and construction techniques, northern communities will become more comfortably livable.

REFERENCES

EPEC Consulting Western Ltd. 1981. *Community Water & Sanitation Services, Northwest Territories.* Report prepared for the Water and Sanitation Section, Department of Local Government of the Northwest Territories, 351p.